An Introduction to the Botany of the Major Crop Plants

Botanical Sciences Series

Editor: J. R. Hillman, *University of Glasgow*

An Introduction to the Botany of Major Crop Plants A. M. M. BERRIE

Plant Steroids B. A. KNIGHTS

Light Reactions and Photosynthesis R. J. COGDELL

An Introduction to the Botany of the Major Crop Plants

Alex M. M. Berrie

Department of Botany
University of Glasgow

London—Bellmawr, N.J.—Rheine

Heyden & Son Ltd., Spectrum House, Alderton Crescent, London NW4 3XX
Heyden & Son Inc., Kor-Center East, Bellmawr, N.J. 08030, U.S.A.
Heyden & Son GmbH, 4440 Rheine/Westf., Münsterstrasse 22, Germany

ISBN 0 85501 220 X

Set by Eta Services (Typesetters) Ltd., Beccles, Suffolk.
Printed in Great Britain by W. & J. Mackay Ltd., Lordswood, Chatham, Kent.

Contents

Foreword

Dr A. M. M. Berrie's book is the first in this Plant Science Series. It is the intention of the Series to provide authoritative accounts of specific areas of plant studies, each volume being written by one or more authors expert in selected topics. Most of the texts in the Series are designed with the undergraduate and graduate student in mind, and they seek to give an introduction in those areas that hitherto have been largely neglected by the written word or teaching course.

The study of plants should not require justification on intellectual and economic grounds, yet there is still a lack of appreciation prevalent even amongst scientists of related disciplines. Paradoxically, in many parts of the world mankind struggles to produce enough food for survival; in such regions botany has basic significance to those without formal education. Sometimes, a lack of realization about the vagaries of food production diminishes the value of learning about plants. Every aspect of botany urgently requires attention if sincerity is to be implied in statements concerning the undesirability of starvation.

Undoubtedly, responsibility for the state of plant studies rests collectively on those of us in its employ. In a vast subject where the languages of mathematics, chemistry and physics are becoming prerequisites in a search for precision, many areas have become esoteric and apparently unimportant. Other topics are distinctly practical but often do not enjoy the respect of the scientific community. The titles in this Series will eventually reflect the diverse nature of botany, although the bias is unashamedly towards the applied part of the subject. We shall attempt to apply a rigorous scientific approach to themes that should occupy teaching time in courses on plants in centres of advanced learning. It is indeed fitting that the series should begin with a book on the major crop species.

J. R. Hillman

Preface

There are many texts which undertake to present the agronomy of the crop plants and, in doing so, consider to varying degrees, the botany of the species. It is difficult to give equal botanical treatment to each of the crop species for much more has been done on the botany of those species of considerable economic importance than on the minor types. In this text this is reflected, perhaps even more so.

Prior to the publication of J. W. Purseglove's *Tropical Crops: Dicotyledons* in 1968 there were few suitable modern texts dealing with the extensive range of crop plants found in the tropics. Those interested in temperate crops were not much better served, and though a little outdated J. Percival's intermediate text on *Agricultural Botany* was hardly equalled. As an *ad hoc* solution I prepared a set of work sheets, which included descriptions of the main crop families, for my class in Agricultural Botany. The originals, with revision, have been used for ten years and I was persuaded to make further revision, increase the content, and provide illustrations. The result of this persuasion is the present text.

Since it developed from work sheets the presentation is somewhat dogmatic and while such, the statements are not definitive. In the areas of morphology, anatomy, and systematics the specialist appreciates the development of an interpretation, or the delegation to a particular taxon by means of logical argument, but I have found that the junior student seems to come to an understanding more rapidly if the presentation is unequivocal. That has been the approach here, but the reader should be conscious that in addition to errors of omission and commission any text can err by the presentation leading to misrepresentation. This book will be no exception and I accept responsibility for all such shortcomings.

The order of the chapters reflects the level of commercial importance of the family as determined by an author located on the north-west corner of Europe. Since there has not been any attempt to include a general treatment of plant structure and plant classification this text is intended for students who have completed an elementary course in general botany, and as such the reader can start with any chapter, and read through in any order. Indeed the reading order will be related to any specific course and will depend on the availability of living material.

Any real appreciation of crop plants can only be obtained by the student examining material of the plants in question. In certain climatic zones this demands considerable effort and the teacher will depend on his colleagues in botanical and experimental gardens. My course could not be conducted, and this text would never have been produced, were it not for the contributions made by the staff of the Glasgow Botanic Gardens and the Department of Botany's Experimental Garden of the University of Glasgow. My thanks are also extended to Dr J. R. Hillman who read the text and served as a moderator of my direct approach, and to the large number of remembered and forgotten undergraduates who have made me realize that we know only in part.

Glasgow
November 1976

A. M. M. Berrie

CHAPTER 1

The Gramineae: The Grasses

When we look at the plants around us it would appear that most of them are grasses, or plants that look very much like them. In fact there are some 650 genera, with over 10 000 species in this family and so well represented is it in the world's vegetation that it has given its name to one of the main types of plant community—the grasslands.

If any family can be said to be truly cosmopolitan it is this one with members distributed from the tropics almost to the poles, and from sea level to altitudes as high as plants can grow. This success in distribution is mirrored in the grasses' success in competing with other plants, and it is probably true that there are as many individual grass plants as the others put together. This widespread and abundant distribution implies that the grasses possess characteristics which confer upon them an ability to compete with other plants, to withstand deprivation by foraging animals, and also to be able to survive infection by various parasites.

With many plants the grazing animal is deterred from eating it if the plant has developed spines or bristles. Not many grasses exhibit morphological modifications of this type. Other plants produce poisonous, or bitter, or otherwise distasteful principles. Few grasses are poisonous, although some are bitter. In the absence of such obvious adaptations what is it that has conferred upon grass its capacity to succeed? It is the habit of the plant which gives it this capacity and at the same time a plant form which can be exploited by man to provide him with food and a number of other useful commodities. By husbanding grasses and grassland man can also keep animals able to consume parts of the grass which do not provide him with a food he himself can eat and digest.

Grass fruits are the largest single source of carbohydrate eaten by man. Wheat, rice, maize and sorghum provided Western, Eastern, South American, and African societies with most of their dietary requirement for carbohydrate. With the migrations of man and his cultures, the use of the indigenous grass to the exclusion of all others is no longer found. In Africa maize is as important as sorghum, and wheat now contributes as much to the diet of some Asiatic societies as rice. At one time the grains produced in the temperate zones of the world—wheat, barley, oats, and rye along with rice—were known as the *cereals* while those produced in the tropics and which include maize and sorghum were called the *coarse grains*. This distinction should be thought of as a convenience for traders of these products and not to be based on any botanical differences which might exist. In the temperate regions there is a tendency for the coarse grains to be used as animal feeds.

When the vegetative part of the grass is used to feed animals the species which are grown are called *forage grasses*. These forage grasses can be consumed as they grow, grazed, or harvested and preserved to be used when growth is not occurring in the field because of low temperatures and/or shortage of water. Preservation is either by drying to make hay, or by ensiling. A grass may be dual-purpose, grown either for its grain or as forage. Maize may be grown either for the production of silage or for grain.

The grasses have not only provided us with our major food crops but also with some that are significant in world markets though the product can be considered as non-essential. The major crop which comes into this class is sugar cane. Refined sugar does have some dietary value but because only carbohydrate is ingested; sugar cannot be considered a high quality food stuff. Man's sweet tooth has ensured that there will be a profitable market for this product and in the sugar cane we find the most efficient plant for the production of domestic sugar.

In some regions grasses are grown deliberately for their essential oils (e.g. citronella and oil of vetiver).

The greatest versatility of any of the grasses is shown by the bamboos which, in the tropics, provide structural materials, which can be no more elaborate than the seasoned stem, or processed into planks, gutters, drain pipes and innumerable structural components of houses. Fibres are extracted, but only locally, though some bamboos are used for the production of paper.

To conclude this brief summary of grass products two more items obtained from grains should be mentioned. Maize can have moderately high levels of oil, and as a result of selection there are varieties which contain sufficient to warrant extraction of a highly desirable edible oil much used in cooking and the manufacture of margarine. The other product that may be obtained from the grain is only available after the grain has been fermented. Any grain rich in starch can be used to produce alcohol. In any part of the world we usually find that the local grain is used to give a fermentable liquor when ground and mixed with water. The technique of producing alcohol by distilling the fermented liquor is a relatively recent development in human society and in many societies the distillate is potable, and is called *grain spirit*. Grain spirit which is drunk has an alcohol content of 30–50%. A different type of fermentable liquor is obtained if the grain is allowed to germinate before it is ground and mixed with water. In Scotland the germinated grain is dried before the liquor is made, and this dried germinated material, from barley, is called malt. The distillate eventually obtained from the fermented malt is, after a period of storage in wooden casks, malt whisky. Whole economies rest on this secondary product from the grass. Industrial alcohol is normally produced from cheaper sources of carbohydrate.

If the fermented malt liquor is stored, and treated with materials to curtail unwanted microbial activity, it gives rise to beers, ales and lagers. Barley, rice and sorghum are the grains most often used for beer production but any grain could be the source of the malt.

The annual world production of grain is substantial. In 1974 more than 1.1×10^9 tonnes were produced and one country, the USA, was responsible for about 20% of that total. Sugar cane gives us more than half the world's supply of sugar—about 4.8×10^7 tonnes of a total of 8.0×10^7 tonnes. It is not possible to estimate the world's annual production of forage grasses but it must be higher in gross weight than grain levels. In approximate monetary terms for those grass products which could have reached the commodity markets directly, the value of the grass in 1974 was at least £6.7×10^{10}. Any estimate of the value of all grass products is conjectural but if the last figure were multiplied by three the resulting total may be thought of as a reasonable estimate. It is easy to see that we must understand how a grass develops, and how one grass is related to another, in order to be able to grow the plants well and possibly, bv understanding them better, produce better crops.

The Grass Grain

In farming or commerce that part of the grass most often handled is the so-called seed. This reproductive structure in the grass is in strict botanical terms not a seed but either a fruit, or a fruit further protected by parts of the floral apparatus that remain when the fruit is shed. The botanically non-committal term *grain* should be used for this structure.

The wheat grain is a naked fruit, and indeed the fact that the fruit is not surrounded by inert coverings makes the utilization of wheat easier than it might otherwise be. The grain is called a *corn*, from the Teutonic *korno*, and the word corn is applied to the major indigenous grain of the district. Corn, in England, refers to wheat, in Scotland to oats, in the United States to maize, and so on throughout the English-speaking world. The wheat grain is approximately 7 mm long and 4 mm wide though the grains may be lean or plump with the general form being that of an oblate spheroid. The dorsal side, or back of the grain is smoothly convex but the ventral surface has a deep groove along its length. At the base or proximal end of the grain on the dorsal surface there is a small, distinct patch shaped like a shield. This region contains the *embryo*, and while only comprising about a tenth of the grain, it is the vital component. The remainder is the *endosperm* which is the food reserve used by the young seedling during and after germination. At the top, or distal end, of the grain there is a characteristic mass of long hairs called the brush.

Maize is larger than wheat, being 12 mm long by 6 mm wide and wedge- or tooth-shaped, the narrower end being the proximal part of the grain. There is no ventral groove, and the embryo at the base of the dorsal surface can occupy as much as a third of the whole. There are no hairs on maize.

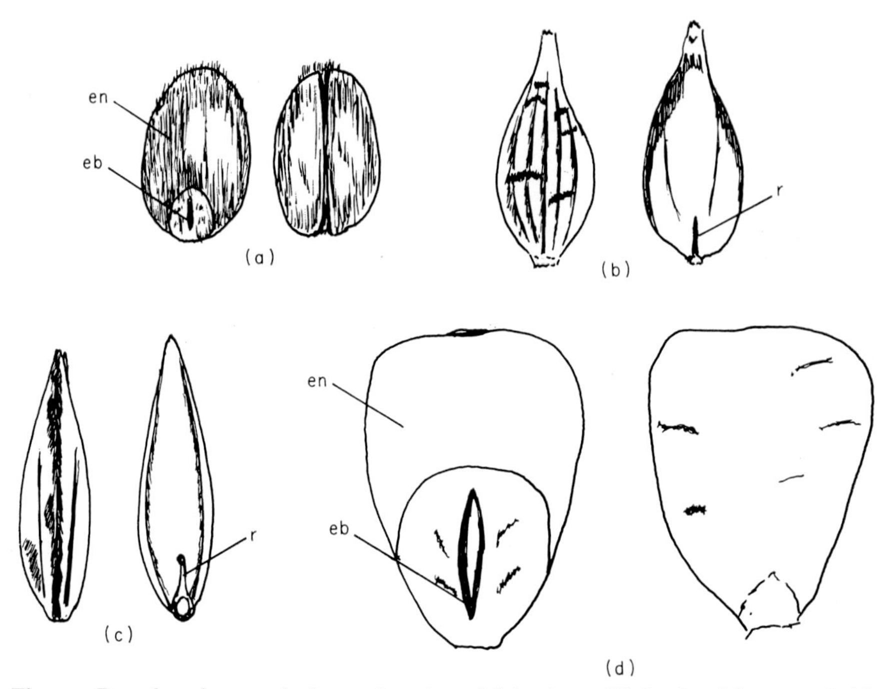

Fig. 1. Dorsal and ventral views of grains of (a) wheat (b) barley (c) oat and (d) maize: eb, embryo; en, endosperm; r, rachilla.

These fruits are characterized by having a single seed and in their development the seed coat, the *testa*, fuses to the fruit wall, the *pericarp*. A fruit of this type with fusion of the testa and pericarp is called a *caryopsis*. All the grasses encountered in temperate zones have caryopses as their fruits but amongst the bamboos we encounter as well as this distinctive fruit, nuts and berries. Bamboos seldom fruit.

Of the other grains rye, and some of the millets, are caryopses, but oats, barley and rice have the caryopsis covered by husks. These are fibrous, with little if any nutritional value, and are usually removed before the grain is milled or prepared for cooking. When the husks are removed the kernel is left. In barley the normal situation is that the husks are fused to the kernel which makes their removal a difficult process, but in the oat the husks are free. When used for animal food or for the production of malt the husks are usually left. The grains of the forage grasses are like those of oat, with a few exceptions.

The dry grain can be stored for lengthy periods and if kept dry and cool wheat grains have been known to remain viable for up to ten years. Even if the viability is lost the grain is still nutritious. When a viable seed is allowed to take up water at normal temperatures it begins to germinate. The first evidence that this is happening is a swelling of the seed and an increase in its moisture content from around 10–12% to 60%.

This embryo is associated with a shield-shaped tissue, the cells of which do not contain starch (Fig. 2). This tissue is called the *scutellum* and the embryo is

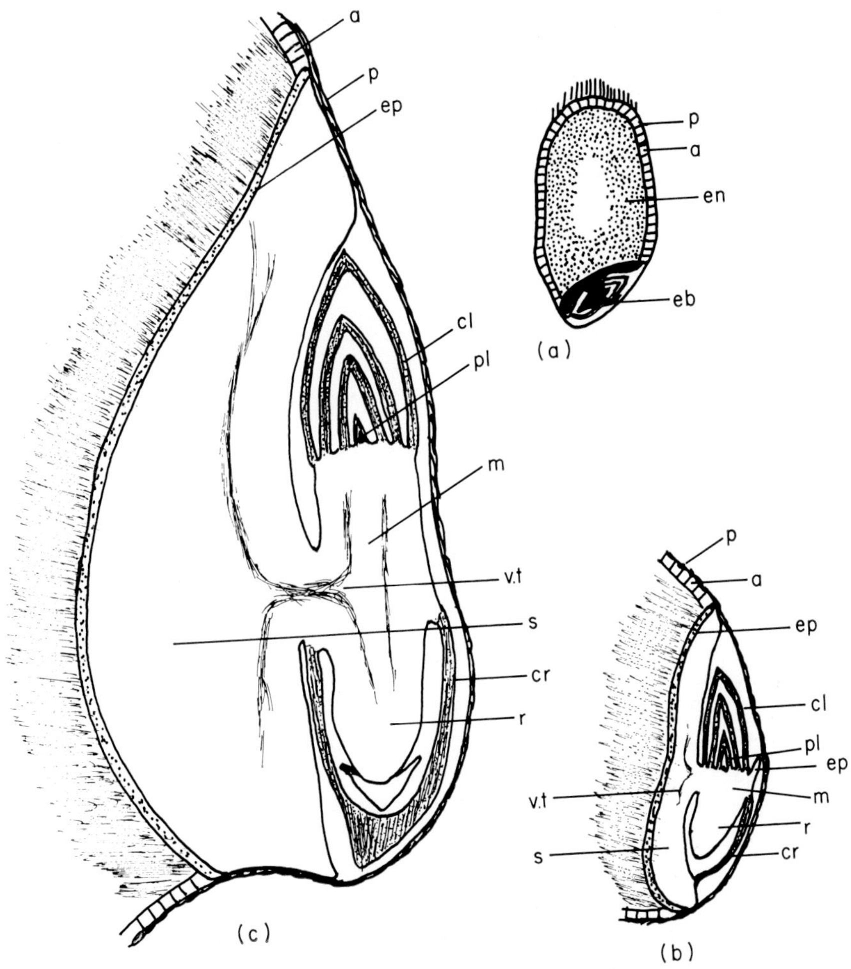

Fig. 2. (a) Longitudinal section of wheat grain: eb, embryo; en, endosperm; a, aleurone layer; p, pericarp + testa. (b) Longitudinal section of wheat embryo: ep, epiblast; cr, coleorhiza; cl, coleoptile; m, mesocotyl; pl, plumular bud; r, radicle; a, aleurone layer; ep, epithelial layer; p, pericarp + testa; vt, vascular traces; s, scutellum. (c) Longitudinal section of embryo of maize; named as for wheat.

attached to it by a peg, arising from its central region. The outer layer of cells of the scutellum where it is in contact with the endosperm are modified and have prominent nuclei and dense cytoplasm. These cells constitute the epithelial layer and it is thought that in the post-germination phase they are concerned with the mobilization of the food reserves from the endosperm to the embryo.

The embryo itself is straight and can be considered as a simple axis. Where the scutellar peg meets the axis it is called the *mesocotyl.* The axis below the mesocotyl, pointing towards the proximal part of the grain is the radicle, which is covered by a protective structure, the *coleorhiza.* Note that in wheat the coleorhiza does not surround the radicle since it is attached to the scutellum at its extreme proximal end. Immediately above the mesocotyl we observe the plumular bud. A sheath surrounds it and this sheath is called the *coleoptile.* The bud itself has a number of leaf initials, usually not more than five. This embryo is quite different from that found in most seeds and attempts to provide homology with the structures usually encountered in the seed have led to controversy over the derivation of the parts. However, it is generally accepted that the scutellum corresponds to the cotyledon readily recognized in other monocotyledonous families.

The wheat embryo has a characteristic structure opposite the point of insertion of the scutellum on the axis. It is a small tongue-like mass of tissue called the *epiblast.* The homology of this is subject to much discussion but if the scutellum is to be considered as a cotyledon this might be thought of as a vestigial second seed leaf.

In wheat there is little development of vascular tissue in the scutellum, but in maize with its larger embryo we can see prominent vascular strands connecting the scutellum and the embryo. But this greater degree of vascularization is not the only difference between these two grains. Maize does not have an epiblast, the coleorhiza extends all the way around the radicle attaching to the scutellum near the mesocotyl attachment. Finally we observe that the plumule is not immediately above the mesocotyl, there being some extension of the axis before the bud proper is encountered.

These differences in embryo features—presence or absence of epiblast, whether the coleorhiza extends all round the radicle, and the extension of the axis at the mesocotyl—are of considerable taxonomic value. One further difference is found between wheat and maize. If the plumule is sectioned transversely the first leaf of wheat has margins which just meet, but in maize these margins overlap. This character, like the other three, is used taxonomically (see p. 22).

If we now turn to the other part of the grain we see that the endosperm consists of a mass of cells densely packed with starch grains (Fig. 3). The nuclei of these starch-filled cells have disappeared during development so we can assume that these cells have no function other than for food storage. Not all endosperm cells are full of starch and enucleate. The exceptions are those on the periphery of this tissue. Usually there is a single row of columnar cells of quite distinct form

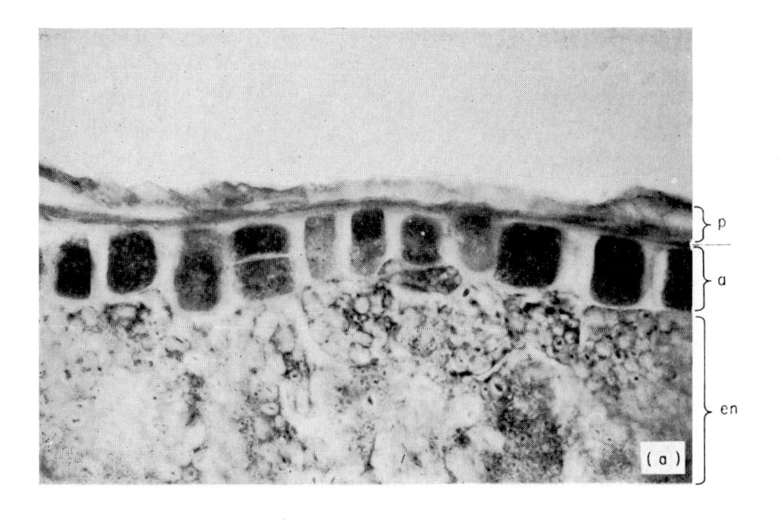

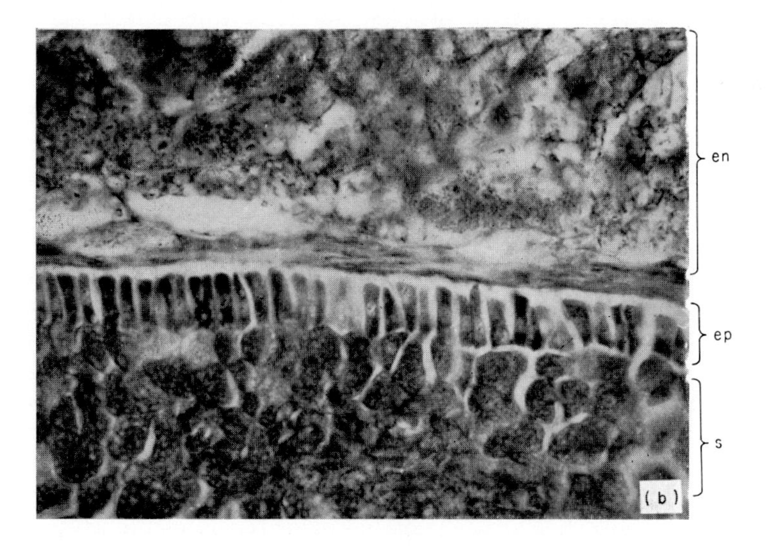

Fig. 3. (a) Section through outer layers of wheat grain: pericarp + testa, p; aleurone layer, a; endosperm, en. (b) Section through inner endosperm of embryo region of wheat: endosperm, en; epithelial layer, ep; scutellum, s. (From a slide prepared by the late J. R. Ashby).

under the coverings of the grain, but very occasionally the layer may be double or even triple in places. The nucleus of these cells is prominent and the cells also contain oil globules and protein crystals which together sometimes make a composite structure which has wrongly been called a crystal. Because of their unique contents we distinguish this peripheral row of endosperm cells as the *aleurone layer* and the protein/oil bodies are called *aleurone grains*.

The embryo and endosperm are covered by the fused testa and pericarp. As examined in sections of the type we have been describing the outer coverings are not well structured. If the grain is broken up and treated with acid and base these layers can be seen in surface view. These cells, along with the starch grains, are characteristic for a species and can be used diagnostically in the identification of the components of a feeding stuff. Starch grains are of two types: simple, ellipsoidal grains with a single growth centre (*hilum*) and with clear growth zones with distinct striae; and compound, in which a number of growth centres constitutes a unit and each centre gives rise to an angular growth without defined striae. Wheat has simple grains in two sizes while maize has compound grains.

When germination is complete, in wheat after about 20 hours at 20 °C, the embryo begins to elongate. At first the coleorhiza extends and ruptures the outer coverings. Almost as soon as these coverings are pierced, the coleorhiza ceases its extension and in turn is penetrated by the radicle. Growth of the coleoptile follows and this structure can enlarge considerably. The pericarp/testa is broken by the growing coleoptile adjacent to the embryo, but if the caryopsis is surrounded by husks the coleoptile often grows between them and the kernel, and does not become evident until it breaks through the distal end of the grain. The coleoptile will continue its growth while it remains in darkness and examples of coleoptiles 20 cm long are known. In some species the region of the axis immediately above the mesocotyl will extend as the coleoptile is growing, for example in oats, while in others e.g. wheat, this region, called the mesocotylary internode, does not begin to extend until the coleoptile has ceased growing. As the coleoptile pushes upwards, it being negatively geotropic, the first leaf in the plumular bud also grows within the coleoptile tube. On breaking the surface of the soil the coleoptile stops extending, due to the stimulus of light, but the first leaf continues to grow to break through the tip of the coleoptile. Other leaves follow, but at this stage no other region of the axis, except in specific instances the mesocotyl, extends.

After a little while the first and second internodes elongate but do not bring their associated nodes above ground. The remainder of the axis is small and the net result of this is that the apical bud is at or about ground level. Leaves will continue to be produced and some slight internode extension occurs but at this stage in development the growing point remains close to the ground (Fig. 4).

The radicle does not remain the only root produced by the developing seedling. Soon after the radicle emerges we find that there grows from the axis a set of roots, each clothed in its own coleorhiza. These roots arise from initials present

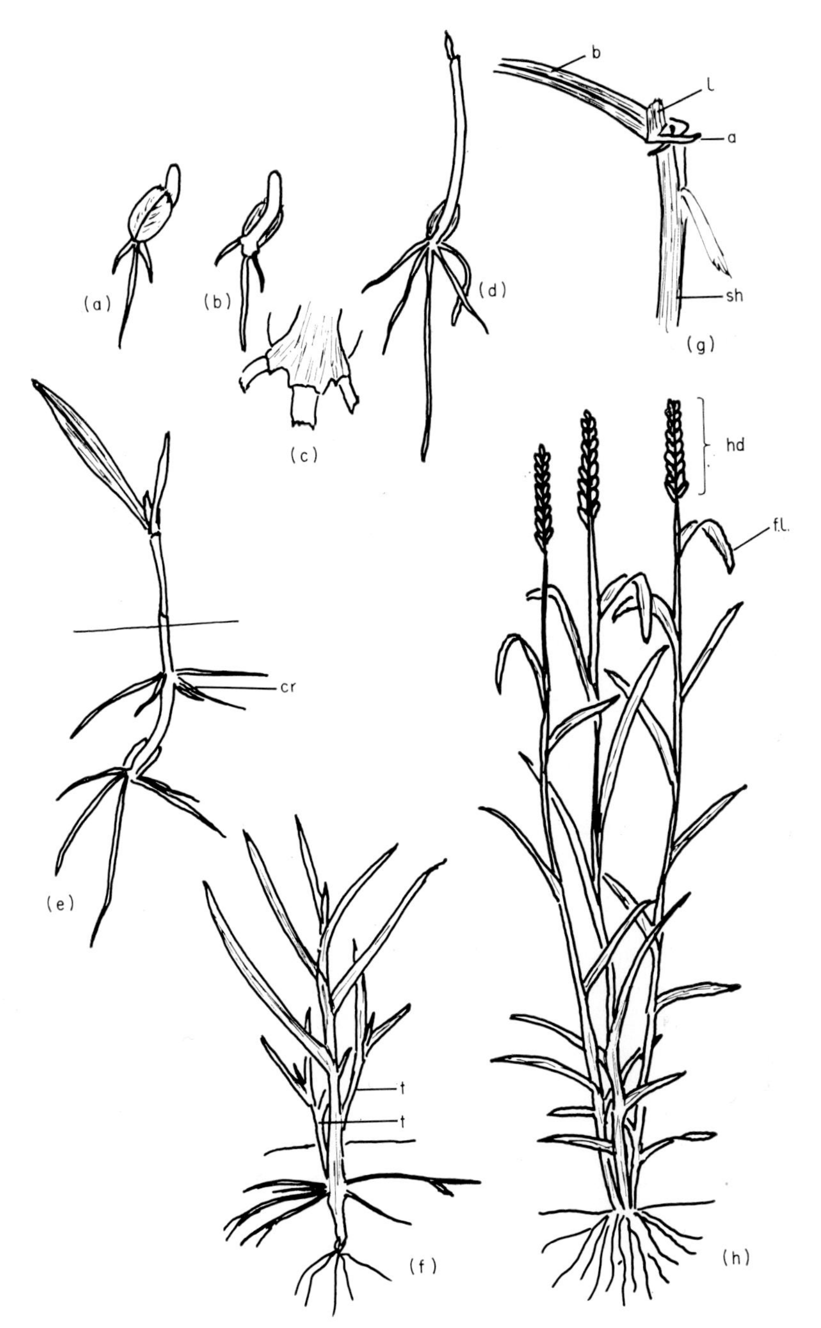

Fig. 4. Diagrammatic representation of the life cycle of a cereal (after wheat) (a) Ventral view. (b) Dorsal view of germinated grain with seminal roots and extended coleoptile. (c) Enlargement of basal region to show coleorhiza surrounding each seminal root. (d) Older seedling with full complement of seminal roots, and tip of first leaf rupturing coleoptile. (e) Expansion of leaves and development of coronal roots, c.r. (f) Tiller production, t. (g) Detail of part of an adult leaf: a, auricles; l, ligule; sh, sheath; (h) Adult plant; hd, head; fl, flag leaf.

in the embryo just below the mesocotyl, and because they are present within the seed they are called *seminal roots* and their number is supposed to be characteristic for the species. While the radicle and seminal roots are well provided with root hairs they do not undergo extensive branching, at least in the early stages. The grass, however, does produce an extensive root system from a set of roots developed from the shoot axis. As such, this second root system is adventitious and it arises from the subterranean nodes. Since this is the crown region of the plant this second root system is made up of *coronal roots*.

This capacity to root at the nodes is kept by the grass throughout its life and it is commonplace to find that whenever a node is in contact with soil, or is buried, adventitious roots arise so extending the root system.

During the period when the grass is growing vegetatively internode extension is limited. New leaves are produced at the rate of one about every five days but the rate of expansion of these leaves may be less. In the axils of the older leaves the axillary buds begin to expand to give branches with the same characteristics as the main axis. We noted how the main axis when it was embryonic was covered by a protective sheath, the coleoptile. The axillary buds in their embryonic stage are likewise covered by a protective sheath, the *prophyll*. The bud grows either vertically upwards, in which case we speak of the development as *intravaginal*, or it can tend more to horizontal growth to give an *extravaginal* pattern. The adult habit of the grass is determined by the growth patterns of the axillary buds, intravaginal growth giving rise to bunch or tufted grasses while the extravaginal growth pattern results in the grass being creeping, producing either stolons above ground or rhizomes below the soil surface. Creeping grasses are most important as soil binders in areas subject to erosion, but some of the worst weed grasses are those with well-developed rhizomes, e.g. couch grass.

The branches which develop are called *tillers*, and the degree and type of tillering helps to determine the grasses' capacity to withstand grazing by animals. It can be seen that the low-set growing points of the main axis and the tillers are well protected from the cow or sheep, but since the latter is able to graze more closely to the ground it is better able to reach some types with growing points at a more vulnerable height. In a grassland with mixed species, any with high growing points might be removed if the grazing animal were a sheep, resulting in a change in the composition of the grassland. We can see that there will be a close relationship between the morphologies of individual grasses, the characteristics of the grazing animal, and the composition of the grassland.

If a grass plant is allowed to develop fully throughout the vegetative stage it can produce a very large number of tillers. Spaced oats growing under very rich conditions have been known to form more than four hundred. Eventually the plant will come into flower but the tillers do not ripen together. This is undesirable because for the successful harvest of a cereal all the heads should ripen at the same time. With cereals the degree of tillering should be at a level to lead to an adequate cover of the ground, to replace losses due to weather, disease and pests, but not such that any plant has a spread of ages amongst its axes. At harvest a single cereal plant should have not more than six tillers if the balance we desire is to be met.

The leaves of the grass during the period of tiller development are protecting the growing points. They do this by their unique form. This unusual leaf is seen to be in two distinct parts, a basal tubular sheath and a terminal flat blade. In the bamboos the leaf is more like that found in other angiosperms, with a short stalk which could be thought of as a modified petiole since there is a small lamina. Grass leaves vary in length from those whose sheaths and blades together measure 10 cm to those whose blades alone are about 1 m long. In width the blades range from about 2 mm to almost 15 cm. The smaller blades are not flat; indeed they are spoken of as solid, and whether a grass has a solid or a flat blade is a useful diagnostic feature. The sheath might be entire or have a split extending from its top towards the base. This is also a useful feature in the identification of grasses.

Where the blade and the sheath meet there are two features best seen in a cereal such as barley. At the transition point from blade to sheath, at the base of the upper surface of the blade, a membranous structure develops. This is the *ligule*, the function of which is thought to be to exclude water from the sheath. In a few grasses, instead of a ligule there is a row of short hairs. Opposite the ligule at the top of the sheath there may develop horny overlapping outgrowths which can be likened to claspers. These are called *auricles* and are seen prominently in grasses with split sheaths.

When the leaf is formed on the axis the primordium expands laterally until it surrounds the stem. From this encircling base the leaf grows up, with the sheath following naturally. The next primordium to be produced is directly opposite the point of inception of the previous leaf and within the developing tube. When the leaves have grown out it can be seen that there are two ranks of them on any axis. When leaves are arranged thus we speak of this as a *distichous arrangement*.

In many grasses the leaf insertion on the stem is swollen to a *pulvinus*, and this can give the node a spurious swollen appearance. The true node is very small but the pulvinus can be large. If the upright stem is displaced bending occurs at the node to re-establish the vertical habit. It is the pulvinus rather than the node proper which is involved in this bending.

Normally after a leaf has been laid down, any increase in size results from cell expansion. The grass leaf has two regions where meristematic activity continues throughout its life: one is at the base of the sheath, the other lies towards the base of the blade. These are the *intercalary meristems* of the leaf. As a result of possessing these zones of growth, if the terminal part of the leaf is removed by being eaten by an animal, the plant can compensate for the loss. So here is another feature that could contribute to the success of the grass under grazing conditions.

Vegetative development is followed by the grass entering a phase of growth associated with reproduction. The stimulus to enter this phase is provided by the environment, and the first expression of it is a change in the form of the apex. In the vegetative stage the apex is a smooth, rather obtuse dome, but after the axis takes on the reproductive form the apex becomes elongated and its tip is more acute. Along the flanks of this elongate apex the flowering head begins to arise, first as a series of bumps (primordia) which then grow around to give a little ridge. In some species this is called the single-ridge stage. Later another primordium arises within the axil of the first ridge to give a double-ridged appearance. From this point on, the apical region's development is determined by the form of the eventual flowering head.

While those events are taking place the lower internodes of the axis are beginning to extend and a stem is produced, with either solid or hollow internodes. This stem is called the *culm.* After identifiable flower initials have been laid down at the apex, culm development is rapid and the young flower head is carried through the encircling leaf sheaths to be exposed. The leaf immediately below the flowering head is seen to have a blade which is shorter and broader than normal. To distinguish it from the others it is called the *flag leaf.*

The Anatomy of the Grass Plant

In its anatomy the grass is a typical monocotyledon as far as the root and the stem are concerned. The root has a central stele which is polyarch and there is a well-developed medulla. The endodermis is prominent, and in the more mature root, passage cells are prominent. The cortex is distinct and the piliferous layer is marked. However, when we look at the piliferous layer in detail we note that there are two main types. The first is distinguished by having alternate cells of the layer produce root hairs arising at an angle from the cell's surface and at a point near the distal end of the cell. In the other arrangement of root hairs, these form from the middle region of the cell and they develop normally (in the geometrical sense) to the surface. The form of the piliferous layer is a good taxonomic character.

The bundles of the stele are scattered throughout the ground tissue in a solid stem, but are restricted to the profile of the stem if it is hollow. Each bundle is well defined with two prominent metaxylem vessels joined by tracheids, while lying generally in an endarch direction is the protoxylem with an associated lysogenous cavity which is thought to carry water in the same way as a vessel. Each bundle is surrounded by a sclerenchymatous sheath (Fig. 5).

The leaf blade has the veins arranged parallel to the long axis and this conforms to the venation expected in a monocotyledon. There is a substantial central region but this should not be considered a true midrib, although it

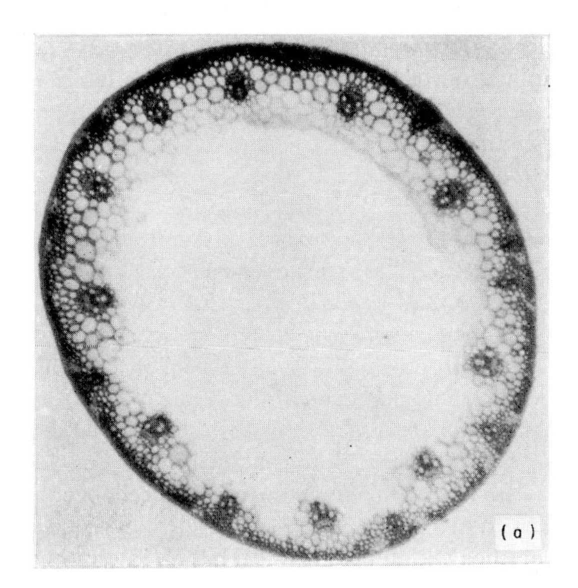

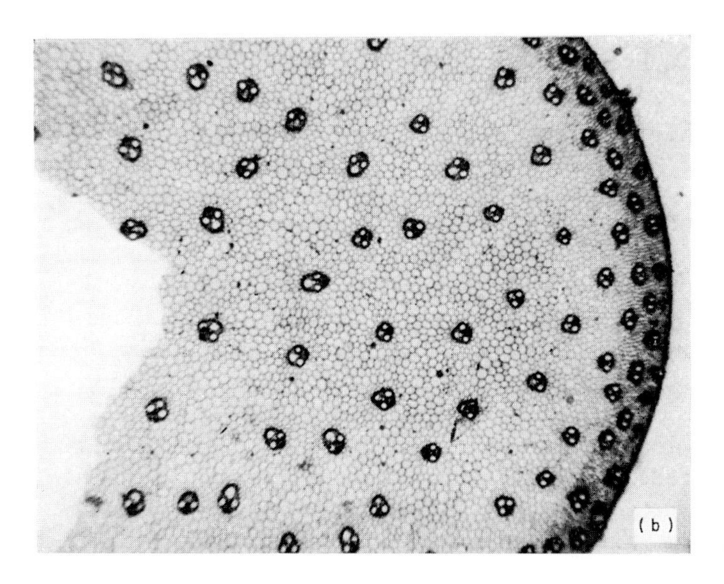

Fig. 5. (a) T.S. of oat internode. (b) T.S. of maize internode.

serves the same mechanical supporting role that we associate with the midrib of the dicotyledon leaf.

A surface examination of the leaf reveals that the blade has a number of unique features. Between the regions of the veins the stomata are arranged in files and these stomata differ from those of all other plants. The guard cells are dumb-bell-shaped, and by changes in the sizes of the ends, the slit between the guard cells is closed or opened. This mechanism is more readily perceived than that of the standard stomatal apparatus. Associated with each guard cell is an accessory, or subsidiary cell. These accessory cells are of two main forms, triangular or flat domes, and again in their shape we have a good taxonomic character. The epidermal cells vary in size and shape but over the veins themselves the cells are small, and at intervals, especially near the margin, there are cells with an opalescent appearance. This is due to the cells having a deposit of silica, and they are opalescent for it has been shown since that the silica in these cells is in the same form as that found in the semi-precious opal. The forms of the silica cells and arrangement of the small cells over the veins are good taxonomic characters.

The presence of silica within the grass leaf has an interesting association with the teeth of herbivorous animals. In some animals the dentition is adapted to allow the teeth to withstand the abrasive grass leaf. It is also well known that if herbivores graze grasses rich in silica the rate of loss of tooth material is high and the animals' lives are shortened.

In transverse section another epidermal cell type is found. These cells form files and may be present only in the mid-region, or regularly over the extent of the blade. In section they are seen to be larger and more thin-walled than the epidermal cells proper, and indeed look like small blisters. Their form gives them their name, *bulliform cells*, though their other name, motor cells, gives a better indication of their function. The size of the motor cell depends on its degree of turgidity and if fully turgid it is a large blister pressing on adjacent cells. The result of this pressure is to keep the blade flat, and in a contrary fashion when the cells are flaccid the adjacent cells are pulled together and the blade is deformed either to be folded about its middle or rolled with one margin inside the other, depending on whether the motor cells are confined to the mid-region or are present as evenly spaced rows. In the unopened bud the leaves show the rolled or folded character and this is a useful character in identifying grasses.

The veins seen in transverse section break the blade into units, for there is usually, between each vein, a region with a system of air spaces. A well-defined palisade is not encountered but the mesophyll is nonetheless distinctive. Again, as we found on the surface of the leaf, two distinct categories can be observed. In one we see around each bundle a distinct sheath of large cells with few chloroplasts. This was first described by the German word Kranz for wreath, and we now know that plants with these prominent bundle sheaths have a different physiology and biochemistry in their photosynthesis from those lacking them. The geographical distribution of the former group tends towards the tropics whereas the plants without this type of bundle sheath tend to be found in the temperate zones. The association of the anatomy, physiology and geographic distribution allows us to characterize these plants as having the Kranz syndrome, and an examination of the leaf anatomy permits the assumption that if the leaf has prominent bundle sheaths it also possesses the distinctive photosynthesis of a Kranz species.

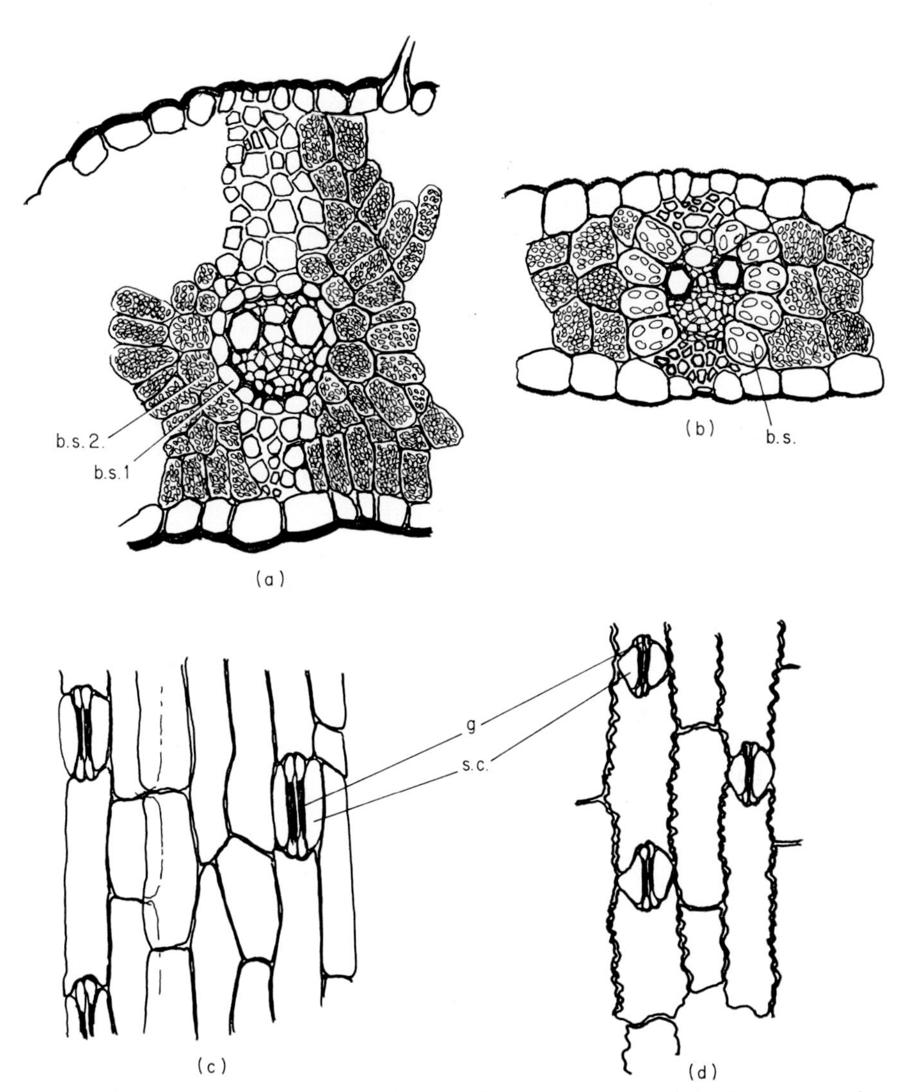

Fig. 6. (a) T.S. of leaf of rye. (b) T.S. of leaf of young maize. (c) Surface preparation of leaf of rye (festucoid type). (d) Surface preparation of leaf of young maize (panicoid type); b.s., bundle sheath; g, guard cell; s.c., subsidiary cell.

The Reproductive Grass Plant

Grass flowers are carried in inflorescences of 1–15 flowers called *spikelets*. Spikelets do not exist singly but are grouped together to form heads and that part of the culm which carries the head is called the *rachis*, on the branches of which the spikelets arise. These secondary axes which come from the rachis and carry the spikelet are the *rachillas*.

Flowering in a grass may be a once-and-for-all event, in which case the grass is *monocarpic*. Perennials as well as annuals can be monocarpic and the Gramineae exemplify these types in a way in which no other family does. The monocarpic perennial is rare but amongst the bamboos it is the rule, and when these large woody grasses flower it is a gregarious phenomenon, all the plants in a region flowering in the same year, the event not re-occurring until the new crop appears and matures, which might be some 20 years later. Annual monocarpic types are well exemplified by the cereals.

Polycarpic forms produce flowers in each of a number of growing seasons and indeed can produce more than one crop of flowers in the same growing season. Forage grasses of the short- and long-term grasslands are of this type, but on the production of flowers the growth potential of the grass is a reduced, which means that the yield from grasslands falls around and just after flowering. By having grasses which flower at different times in the composition of the grassland, production can be equalized over the season, so leading to a more uniform output of animal products such as milk from pasture.

Except in a very few instances, one of which is maize, the spikelet contains at least one flower which has both stamens and a carpel. We might say that the grass spikelet is hermaphrodite. This is hardly correct usage but it serves to emphasize the exceptional morphology of maize which, along with some other members of its tribe, is monoecious. This is because the prominent head of maize is made up of spikelets each of which has two staminate flowers. The carpellate flowers are carried not in heads but on specialized flowering axes that arise in the axils of leaves in the mid-region of the culm.

Individual flowers are naked since the perianth is not present as sepals and petals. The protective function of the perianth is taken over by the enlargement and specialization of two structures which can be thought of as homologous to bracteoles. The group of flowers, each with associated bracteoles, is in turn subtended by two bracts. At the base of the flowering axis there is the first of the modified bracts, and this is called the lower glume. Above it on the rachilla there is a similar structure, the upper glume. In size the glumes may vary from as small as a bristle-like structure, e.g. barley, to being so large that they enclose the whole spikelet, e.g. oat. Glumes can be cornaceous, papery or delicate, and the vascular bundles, which correspond to those of a leaf, are retained. The central bundle can form a distinct keel, about which the glume is more or less folded, e.g. as in wheat, the lateral bundles making up the nerves. Since the number of nerves is usually a constant character, this is useful in the identification of grasses. The upper glume lies directly opposite the lower and as we continue distally along the rachilla we see that the lowermost (oldest) flower is directly opposite the upper glume.

The flower is inserted on the rachilla, there not being a prominent axis on which the bracteoles and the essential floral parts arise. The lowermost of the bracteoles is termed the *lemma*, or lower pale, and nearly always possesses a keel and has distinct nerves. On the keel there may or may not be unusual substantial bristle-like structures. When they are present, these structures, *awns*, can arise from the apex of the lemma in which case they are apical awns, or from just below the apex (sub-apical awns), in the middle of the keel (dorsal awns) or towards the base (basal awns). Dorsal and basal awns are different in character than those arising from the tip region. In the earlier stages of development of the flower, tip awns contain chlorophyll and in the very large awns of barley stomata can be discerned. Certainly in the case of barleys the awns are actively transpiring and producing carbohydrate by photosynthesis, and we know that the activity of the awn is responsible for some of the seeds' food reserves. Dorsal and basal awns are not green and do not have functional stomata. Often these awns have specialized spiral thickenings at the lower region and are sharply bent above this. With changes in atmospheric moisture levels the spiral thickenings twist and untwist and in high moisture conditions the sharp bend can straighten. This type of awn functions in seed dispersal, and in the case of wild oat can bury the seed very rapidly.

The second of the bracteoles does not have a prominent central keel and is termed the *palea*, or upper pale. Up until anthesis the lemma and palea enfold the essential parts of the flower. At the time of anthesis the pales are forced apart by two (in some species three) very delicate organs called *lodicules*. More than one interpretation exists regarding the homology of lodicules but it is generally accepted that they are equivalent to perianth members. Though the lodicule is delicate and membranous, at anthesis it becomes gorged with water very rapidly, and forces the pales apart to expose the stamens and carpel (Fig. 7). The filaments are also taking up water at the same time as the lodicules and this additional increase in bulk of organs within the pale cavity forces them still farther and faster apart. In favourable conditions it is possible to watch a filament extend from less than one to over five millimetres in almost as many minutes.

Three stamens are exerted at anthesis, but six are present in the flower of rice, and this larger number prevails in the bamboos. Fewer than three stamens can occur but this reduced situation is normally a generic feature, e.g. one stamen in *Vulpia*. The stamens are large with prominent anthers on long filaments inserted into the middle of the connective, giving a versatile anther. Dehiscence is by apical pores, and there is prolific production of pollen.

Surmounting the floral axis there is a single carpel with two, or rarely three feathery stigmas. There is a single basal anatropous ovule.

Pollination is by wind and it can be seen how well adapted this particular floral mechanism is for anemophily. Self-pollination is not excluded unless pollen is ripe at a time when the stigmatic surfaces are non-receptive. *Protogyny* does occur as does *protandry*; however, even in the absence of these, most grasses show a high degree of cross-pollination with consequent cross-fertilization. Notable exceptions to this predominantly outbreeding situation are found in the cereals wheat, barley, and oats. Indeed, in these cereals the pollen is often shed before the pales open and self-pollination is obligatory, almost establishing cleistogamy.

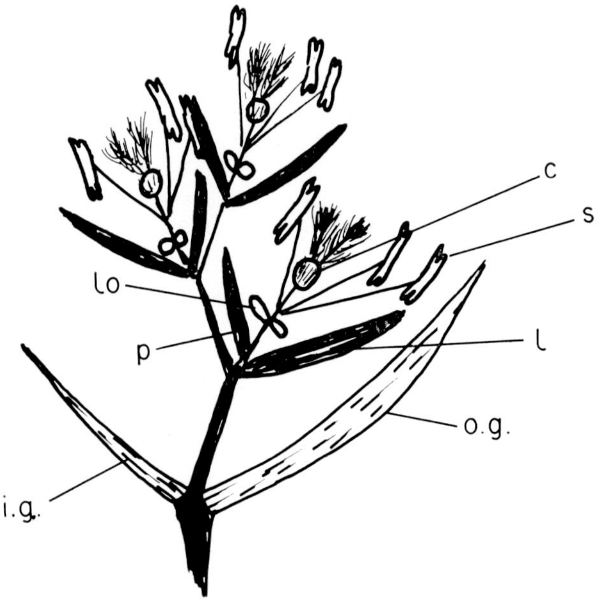

Fig. 7. Diagram of grass spikelet: o.g., outer glume; i.g., inner glume; l, lemma; p, palea; lo, lodicules; s, stamens; c, carpel.

We consider that the primitive grass had a spikelet that was many-flowered. If the general assumption that reduction in numbers of parts represents advancement then there are two ways the number of flowers in the spikelet can be lessened: from the base upwards (basipetally), or from the apex downwards (acropetally). Both types of reduction are found.

As remarked, the flowering head of the grass is a compound inflorescence. Spikelets can be carried on rachillas which are distant from the rachis, to give an open head called the *panicle*, or they can be sessile and opposite on the rachis, giving the *spike* (Fig. 8). In spikes the spikelets can be diametrically opposed, or they are opposite by an angle of about 60°. In the latter case the head is made up of a system of spike branches. The panicle is exemplified by oat and the spike by wheat. An intermediate condition is found in many species where the head is a panicle but the spikelets are not carried distant from the rachis but

Fig. 8. Diagrams of various forms of grass heads. (a) True spike with spikelets sessile, opposite and alternate. (b) Branched spike with spikelets sessile, and at approx. 60° from each other. (c) Panicle. (d) Spike-like panicle, spikelets stalked and spirally arranged: r, rachis; rl, rachilla.

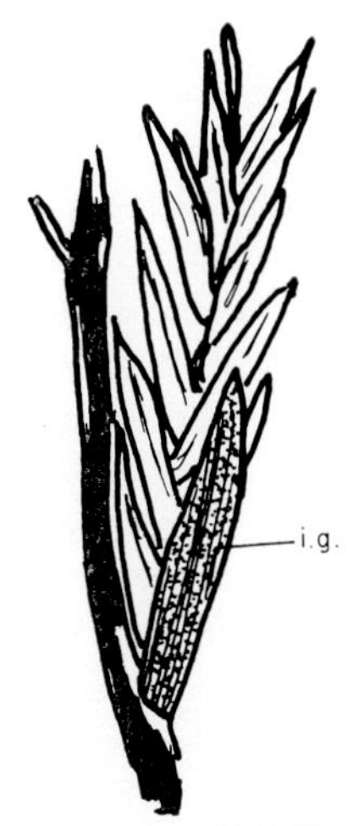

Fig. 9. Spikelets of *Lolium perenne* showing radial disposition and outer glume adnate to rachis: i.g., inner glume.

are on much-contracted rachillas. Such an arrangement gives a head which is superficially like a spike and we term it a *spike-like panicle*. It is distinguished from a true spike by the spikelets not being opposite or sessile on the rachis. Finally with respect to the form of heads there is the *pseudo-spike* of the rye grasses (see p. 38 and Fig. 9).

Grass Systematics

The taxonomy of the grasses was at first based on the form and character of the spikelet and the great subdivision of the family into its two major subfamilies can still be made without reference to other parts of the plant. But as is nearly always the case a few of the taxa below the level of the subfamily, i.e. tribes, could not be unequivocally allocated to either of the main subfamilies.

It was Robert Brown who considered that the family was constituted of the subfamilies Paniceae and Poaceae, which in conformation with the rules of nomenclature would today be called Panicoideae, and Pooideae, and that these taxa were distinguished on spikelet features and geographical distribution. The Panicoideae are found predominantly in the tropics while the other subfamily is mainly native to the temperate zone. Brown's Pooideae is today that subfamily called the Festucoideae and we can no longer define it by its geographical distribution with the same confidence as before. However, the spikelet characters are still most valuable in separating the subfamilies. The spikelet of the festucoid grass is one- to many-flowered and where there is reduction from the many-flowered state this takes place from the apex downwards. This means that the rachilla is prolonged beyond the terminal flower, and even in the case of single-flowered spikelets, the grain that results is found to possess a rachilla. Panicoid grasses have spikelets with one, sometimes two flowers, but in the latter case the lower flower is imperfect. The basal position of the reduced flower implies that the direction of reduction in this spikelet form is from the base upwards, and to substantiate this the rachilla does not project beyond the terminal flower. Another distinct difference between the spikelets of the subfamilies is that there is no measurable internode between the glumes and pales of the panicoid grass, whereas in the festucoid there is. This particular feature may have a bearing on the other character which is used to separate these taxa, and that is the point(s) at which disarticulation takes place along the rachis. In the panicoid there is only one point at which a break occurs on the ripe rachilla—below the glumes. This means that the spikelet falls as a unit. With Festucoid grasses, breaks occur all along the rachilla at the bases of the individual flowers. Individual grains result, each with a portion of the rachilla present at the base.

There were one or two of the tribes originally assigned to either the Panicoideae or the Festucoideae which seemed not to possess the necessary distinctness of features for taxonomists to be certain of their position.

The original tribal arrangement is as follows:

Subfamily Panicoideae	*Subfamily Festucoideae*
Maydeae	Phalarideae
Andropogoneae	Agrostideae
Zoysiae	Aveneae
Tristegineae	Festuceae
Paniceae	Chlorideae
Oryzeae	Hordeae
	Bambuseae

Bentham used the form of the head, the number of flowers per spikelet, the form of reduction, and the relative lengths of the glumes as the main feature in erecting a tribe. However, four of the tribes were difficult to assign unequivocally to a particular subfamily: Oryzeae, Phalarideae, Chlorideae and Bambuseae. Bentham, and later 19th century taxonomists accepted that the Oryzeae and the Bambuseae were indeterminate; indeed that is why they are listed last, but it was not until later that the other two were shown to be sufficiently unusual to present exceptional cases. With the Phalarideae it is of interest to note that the early taxonomists mistook the two lowermost reduced flowers of a three-flowered spikelet as another pair of glumes. Now that we know the homology of these structures we can recognize that there is a close affinity with the Aveneae, and the genera of the Phalarideae are now included in the oat tribe.

From about 1890 until 1930 a vast amount of factual material relating to the structure of grasses was accumulated, and some of this strengthened the belief in the difficult positions of the four tribes just mentioned. Since 1930 the taxonomy of the Gramineae has evolved by collating and using the morphological details amassed over the preceding years, and incorporating new information as it was discovered. We still consider the Festucoideae and Panicoideae as the main subfamilies and any particular part of the grass plant is stated to possess festucoid or panicoid features. When all the features are of the same category there is no difficulty in assigning a grass to its subfamily. It is either a member of the Festucoideae or the Panicoideae. When, however, the features straddle the categories, assignment of the individual to a subfamily is difficult. As the information has been collated it has been found that there is a consistency in the way the intermediate mixtures of characters festucoid and panicoid occur, so much so that new distinct subfamilies have been able to be established. Six such subfamilies are now recognized: Arundinoideae, Bambusoideae, Eragrostoideae (Chloridoideae), Festucoideae, Panicoideae, and Oryzoideae.

The characters which are used as panicoid and festucoid are listed in Table 1.

The 13 tribes of the original classification of Bentham have been increased to over 40, but in this area of grass systematics personal opinion often dictates the establishment of a distinct tribe. It is generally agreed that there is a need for

Table 1

A comparison of the features used in the separation of the major subfamilies of the Gramineae

Panicoid characters	Festucoid characters
Caryopsis	
Embryo + scutellum about one-third of whole.	Embryo + scutellum not more than one-quarter of whole.
Embryo	
Vascularization prominent, traces to coleoptile distant from mesocotyl.	Vascularization slight, traces to coleoptile at same level as mesocotyl.
Epiblast absent.	Epiblast present.
Coleorhiza attached to scutellum near mesocotyl.	Coleorhiza attached to scutellum at proximal end.
Margins of first leaf overlapping.	Margins of first leaf not overlapping.
Any embryo is designated as P − PP if all the features agree with those listed for a panicoid grass and F + FF if festucoid. Intermediates may be designated as P − PF meaning the first three features are Panicoid but the last one is Festucoid (see e.g. Arundinoideae, below).	
Seedling	
Blade of first leaf broad, about one-fifth length.	Blade of first leaf narrow, about one-ninth length.
Leaf surface	
Stomatal subsidiary cells triangular, or pronounced dome-shaped.	Stomatal subsidiary cell roughly rectangular, or low-domed.
Bicellular hairs frequently present.	Bicellular micro-hairs absent.
Silica cell saddle- or dumb-bell shaped.	Silica cells elliptical or crescent-shaped.
Short cells over veins in pairs.	Short cells over veins in rows of more than five.

The characters used to designate the remaining subfamilies are as follows:

Arundinoideae

Embryo: either P − PF or P + PP.
Leaf surface: both festucoid and panicoid characters present.
Leaf interior: festucoid.
Spikelet: festucoid.
Roots: panicoid.
Chromosomes: panicoid.

In this subfamily the ligule is not a simple membranous flap. It is either a small, fringed membranous structure or has been replaced by a ring of hairs.

Bambusoideae

Embryo: F + PP.
Leaf surface: both festucoid and panicoid characters present.
Leaf interior: festucoid.
Inflorescence: festucoid, but flower with three lodicules and often six stamens.
Roots: panicoid.
Chromosomes: panicoid affinities.

Table 1—*contd.*

Panicoid characters	Festucoid characters
Leaf interior	
Bundle sheath often of large chlorenchyma with specialized chloroplasts.	Bundle sheath endodermal in character.
Inflorescence	
Spikelet with one fertile flower and sometimes a second basal reduced.	Spikelets with one to many flowers and, if reduced, the upper flowers usually missing.
Disarticulates below the glumes.	Disarticulates along the rachilla above the glumes (exceptions occur).
Lodicules usually small and thick.	Lodicules usually long-pointed and membranous.
Roots	
Piliferous layer cells all alike.	Piliferous layer cells alternating large and small with root hairs.
Root hairs arising at middle of cell and more or less at right angles.	Root hairs arising at distal end of cell and pointing towards root tip.
Some authors include other characters such as the nature of the food reserves and whether the culm has solid or hollow internodes. These are not reliable characters and are best left out of our consideration.	
Physiology	
Many species exhibit C4 type photosynthesis.	Only C3 type photosynthesis present.
Many species germinate in the presence of 0.5% Monuron (a herbicide).	Germination inhibited by 0.5% Monuron.
Low oxygen tensions not inhibitory to germination.	Oxygen tensions below 3% inhibit germination.
Warm climate plants often short day requiring and not having dormancy broken by low temperatures.	Cool climate plants requiring long days for flowering and dormancy broken by low temperatures.
Cytology	
Basic chromosome number 9–10, mitotic chromosomes small.	Basic chromosome number 7, mitotic chromosomes large.

Eragrostoideae

Embryo: P+PF.
Leaf surface: mainly panicoid.
Leaf interior: mainly panicoid; some with very prominent bundle sheath.
Inflorescence: festicoid-like.
Roots: panicoid.
Chromosomes: panicoid affinities.

Oryzoideae

Embryo: F+FP or F+PP.
Leaf surface: mainly panicoid-like but sometimes festucoid characters are seen, e.g. small cell over veins in rows more than five.
Leaf interior: festucoid.
Inflorescence: characteristic with a single flower which may have six stamens.
Roots: apparently mixed.
Chromosomes: Panicoid affinities.

this large number of tribes to allow us to interpret how groups of genera are mutually related.

Not all tribes contribute to the economically important grasses and in our treatment the less important will not be considered.

The Subfamily Festucoideae

The tribe Hordeae

Undoubtedly the most important tribe is the Hordeae. Some authors call this tribe the Triticeae, but the name Hordeae should have priority. It might be as convenient to call this taxon the cereal tribe, for from its members there have been selected three of the world's most important cereal crops—wheat, barley and rye.

The main distinguishing feature of the members of the Hordeae is the form of the flowering head. It is a spike in which the spikelets are alternating and directly opposed. Generally the spikelets are many-flowered but in the barleys, the genus *Hordeum*, the spikelets are one-flowered. Within the tribe we find two main divisions depending on whether there is only one spikelet inserted at each point of the rachis carrying inflorescences, or two or more.

The main genera of this tribe are:

Triticum L.	the wheats
Hordeum L.	the barleys
Secale L.	rye
Agropyron Gaertn.	the couch grasses or wheat grasses
Elymus L.	the lyme grasses
Aegilops L.	the goat grasses

Other genera of some importance are *Haynaldia* Schur., *Hystrix* Moench., and *Sitanion* Raf. Hybridization between genera takes place indicating the close genetic affinity existing amongst the members of the tribe.

The genus *Triticum*

As wheat has evolved so has human society. All evidence points to the wheats having a centre of origin somewhere in the fertile crescent running from Turkish Thrace through to Afghanistan and including the valleys of the Tigris and Euphrates—that region we call the cradle of civilization. There is evidence that man was utilizing wheat some 10 000 years ago but that was a type that is no longer considered of commercial value.

There are three main groups of wheat recognized primarily on the number of grains that each spikelet produces and on the quality of flour that can be made from the endosperm. There is Einkorn which gives but one grain per spikelet,

Emmer giving two, and Dinkel or Bread with more than two, and sometimes as many as five, grains per ripe spikelet. It is only in the last group that a flour giving a risen loaf (leavened bread) can be produced, Einkorn giving rise to a product more akin to a meal, while in the Emmers the flour gives on baking a product more like a cake, or in the particular case of one species, pastas such as spaghetti or macaroni.

There are few wild species of wheat, and they are recognized because of their extreme hairiness. It seems as man selected his wheats he removed this character and the cultivated wheats, while still hairy, are not prominently so. Also he has selected for easy and uniform germination. This allowed predictability of sowing and reaping.

Triticum is recognized by having large tough glumes with a projecting tip to the keel and at least in the upper part of the glume having a sharply keeled back. Glumes of Einkorn have a distinct lateral tooth, which becomes less pronounced in the case of the Emmers, and with the Bread wheats the lateral region of the glume tip has a rounded shoulder. In most wheats the lemma has a pronounced terminal awn, though certain varieties of Bread wheat are awnless or beardless. The selection of beardless types was dictated during the era when wheat growing was expanding to meet the increase in world population during the 19th century but the crop was still being harvested by hand; the rough awn is most damaging to hands and arms. Since it is known that long green awns can be valuable contributors to grain development, bearded wheats are now becoming more important once again since the combine harvester feels neither pain nor discomfort. The palea is awnless.

All species of *Triticum* have spikelets with many flowers, but as remarked not all flowers give rise to a grain, and almost always the uppermost flowers of the spikelet are sterile. The flowers are typical. At maturity the grains are the free caryopsis but in many species, and always in the wild ones, the rachis breaks up between the spikelets so that the unit that is dispersed is not a grain but a whole spikelet. Such a rachis is called brittle in contrast to the tough rachis that remains entire. Wheats which have a brittle rachis are said to be spelt wheats and the product obtained on threshing is called Vesen. Vesen, the caryopses enclosed in pales and glumes with a portion of rachis attached, can be fed to animals and surprisingly is not much less less nutritious than the caryopses themselves. Vesen is unacceptable to man. The mutation that gave rise to toughness of the rachis has been said to have been the most important in the development of the *Triticums* as crop plants for it is this which allows the whole ripe head to be harvested and then threshed with the caryopsis falling free.

In its vegetative stage the wheats are examples of an annual bunch grass that tillers fairly freely. When grown at crop densities tillering of individual plants is restricted and at harvest it is unusual for any one plant to have more than five to seven tillers. By planting closely and restricting tillering all flowering heads come to ripeness at the same time so ensuring easy harvesting. Two physiological types of wheat exist: one behaves as a winter annual, the other as a spring annual. Winter annual types which are sown in late September or October, in the North Temperate zone, germinate in autumn and over-winter as vegetative plants. During this phase they tend to tiller abundantly and are semi-prostrate. Spring annuals are sown from March onwards and have a much more curtailed period of vegetative growth so do not produce as many tillers, and are upright. Both winter and spring types are long-day plants but the winter wheats will only respond to the appropriate day length after an exposure to low temperatures, i.e. they require to be *vernalized.*

As with all cereals, yield will be conditional on a number of morphological characteristics:

1. The number of flowering heads per unit area of land.
2. The number of spikelets per head.
3. The number of flowers per spikelet which ripen a grain.
4. The weight of an individual grain.

Nutritional levels are of course important and some degree of compensation can take place, so that a variety with an inherent capacity to produce many spikelets per head may if grown in poor conditions produce lower than average grain weights, so reducing the final yield.

Since the potential expression of each of these morphological characters except the first is determined genetically the breeder can select them in the expectation that there will be improvement in yield if (2), (3) and (4) are increased. The eventual yield will be determined by environmental conditions interacting with the genotype.

Other aspects of the plant's morphology and physiology which can affect yield come into play during the time that the grain is developing. We know that the carbohydrate laid down during grain development is produced by current photosynthetic activity and does not come from material produced and stored during an earlier stage in the plant's life. By using radioactive tracers, shading or removal of parts of the plant, it is possible to determine the extent of the contribution of any one part to the amount of carbohydrate that is laid down in the grain. In wheat the parts that produce grain substance are the penultimate leaf on the culm, the flag leaf, the green upper culm, and the floral parts while still green. About 45% of the total is produced by the activities of the green part of the head so it is most important to ensure that there is as much photosynthesizing tissue in this region as is possible, and if the awns are included then there is the possibility that yields could be increased. The contribution made by the two leaves is something which is conditional on the latitude at which a particular variety is grown. These leaves will intercept sunlight most efficiently when the incident solar radiation is perpendicular to their surface. The angle the leaf makes with the vertical culm will be responsible for the establishment of this condition in relation to the sun's azimuth (which of course will be averaged over the days during grain formation). This feature of the flag leaf is one which can be subject to selection because of the correlation which exists between leaf angle and photosynthetic capacity at any particular latitude.

Total weight of grain is not the only aspect of yield that must be considered. Samples of grain either from the different species or different varieties of the same species, or the same variety grown in different places, produce flours which vary in quality and character. As already remarked the flour produced from Einkorn or Emmer wheat is not able to be used in the production of leavened bread, and in miller's terms is considered soft, but in contradistinction not all bread wheat will produce a hard flour capable of giving a dough which will rise.

Except in the case of specially bred varieties, wheat grown in the United Kingdom gives a flour suitable more for the baking of biscuits than bread. This suggests that the climate of the U.K. does not allow the wheat endosperm to build up in the same way as the endosperm of a wheat developing in a drier, hotter climate. Also the high levels of fertilizer used in intensive agricultural systems, and the time of application of fertilizer can determine the chemical composition of the grain and so affect the type of flour that is made.

The origin of modern wheat

Hybridization studies involving members of the three groups of wheat, and the genus *Aegilops* have allowed us to construct a possible evolutionary pathway followed by the genus *Triticum* over the last 10 000 years. Since wheat during this time has been closely associated with man the evolution has been accelerated by the conscious selection of types possessing desirable characters.

A fuller understanding of the way in which the genus has evolved becomes possible when the hybridization studies are coupled with cytological investigations. A hybrid between an Einkorn and an Emmer wheat is sterile and the reason for this is apparent when a count of the chromosomes of the parents is made. The Einkorn parent is found to have 14 chromosomes and the Emmer twenty-eight. The hybrid has 21 and because of unequal distribution of chromosomes to the gametes cannot, or very very exceptionally can, produce viable pollen or eggs. When a chromosome count of a bread wheat is made it is found to be forty-two. The basic chromosome number, x, in the genus is 7 and we see that the Einkorn wheat is $2x = 14$ and is diploid, the Emmer is $4x = 28$ and is tetraploid and the bread wheat with $6x = 42$ chromosomes is hexaploid. The chromosome number allows us now to allocate a given *Triticum* species more accurately to one of our original groups than the morphological characters mentioned earlier. When this is done and associated with the possession of a tough rachis we arrive at a compilation of the wheats that begins to indicate their phylogeny (Table 2).

If an examination of chromosome pairing at meiosis in the 21 chromosome hybrid (triploid) is made it is seen the chromosomes are associated as seven pairs with seven singles left unpaired. This is designated $7'' + 7'$, or in words,

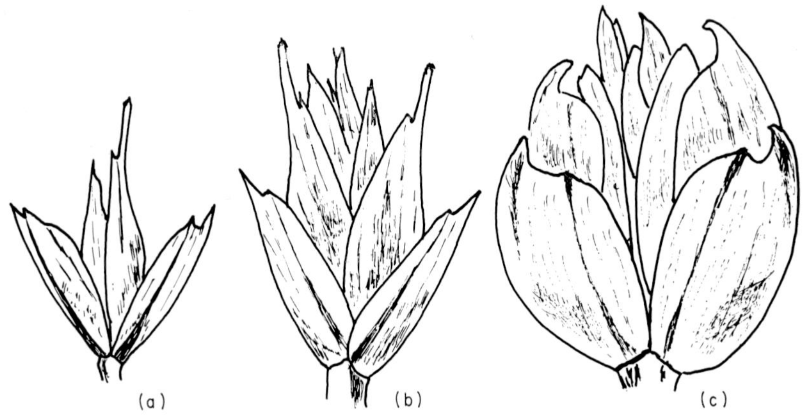

Fig. 10. Spikelets of (a) *Triticum monococcum*, (b) *T. dicoccum* and (c) *T. aestivum*.

seven bivalents plus seven univalents. It can be concluded from this meiotic arrangement that the seven chromosomes contributed by the Einkorn parent are sufficiently like seven possessed by the Emmer to pair with them and this implies that a wheat like *T. monococcum* was a progenitor of *T. dicoccum*. The complement of genetic material making up the basic set is called the genome. The diploid plant's cells each contain two identical genomes and in *T. monococcum* the genome is designated A so the cells of the diploid are each AA and we can conclude that the cells of the triploid are each AA*x* where *x* has been contributed by some other species in perhaps another genus. By carrying out hybridizations between *T. dicoccum* and diploids from related genera and then finding which hybrid had a meiotic configuration of $7'' + 7'$ the origin of genome *x* could be ascertained.

While cytogenetical studies on the hybrid between *T. monococcum* and the Emmer wheats indicated unequivocally the origin of the A genome in the latter, hybridization studies between Emmer and other diploids to determine the

Table 2

The main species of wheat arranged according to cytological group and rachis character

	Brittle Rachis (Spelt)	Tough Rachis
Diploid	*T. boeticum* Boiss. *T. monococcum* L.	
Tetraploid	*T. dicoccoides* Koern. *T. dicoccum* Schubb. *T. timopheevi* Zukov.	*T. durum* Desf. *T. turgidum* L. *T. turanicum* Jakubz. *T. polonicum* L. *T. carthlicum* Nevski.
Hexaploid	*T. spelta* L. *T. macha* Dek. and Men. *T. vavilovi* Tuman.	*T. aestivum* L. *T. compactum* Host. *T. sphaerococcum* Perc.

origin of *x*, mainly from amongst the members of the genus *Aegilops*, were inconclusive. Early work in attempting to determine the parentage of Emmer wheat by comparing the morphologies of suspected parents with the tetraploids suggested that one parent might be *Aegilops speltoides* Tausch. Recently doubt has been cast on this by comparing the constitution of seed proteins from a number of diploids, and tetraploids. The results of this work lead to the conclusion that the B genome chromosomes did not come from a species of *Aegilops*. This genome appears to have been derived from another wheat *Triticum urartu* Tum. This wheat was considered a rare endemic but as a result of a recent expedition it has been found throughout the range of distribution of the wild tetraploid whcats, and there is also found in the same regions the wild diploid wheat *T. boeticum*.

A hybridization programme between *T. boeticum* and *T. urartu* is beginning to provide substantial evidence for the supposition that the B genome of the tetraploids is indeed derived from *T. urartu*. The hybrid itself looks like an Emmer wheat and the chromosome pairing arrangements agree much more with expectation.

A similar programme of hybridization involving the hexaploid wheats showed that their progenitors were a tetraploid, perhaps both *T. dicoccum* and *T. dicoccoides* separately and at different times, and another *Aegilops*, *Ae. squarrosa* L. whose genomic constitution is DD so making the hexaploid bread wheat AABBDD.

The wheats then form a polyploid series in which alloploidy is well established. One reason for the precision of pairing of homologous chromosomes of a particular genome and the avoidance of mispairing with the corresponding chromosome from a very closely related genome, say the second chromosome of genome A with the second chromosome of genome B, so-called homeologous chromosomes, is the occurrence of a gene to prevent this on chromosome 5 of the B genome. This gene has stabilized the hexaploid.

Our conclusions regarding the progenitors of the tetraploid and hexaploid wheats are conjectural and the *Aegilops* species mentioned is the one thought most likely to have been involved, but because of the close genetic relationship existing between *Triticum* and *Aegilops* it is possible to introduce chromosomes and parts of chromosomes from other *Aegilops* species into the wheats. This was done in the wheat variety Compair which has genetic material from *Ae. comosa* Sibth. et Sm. This capacity for a relatively free exchange of genetic material has led some authors to include *Aegilops* under *Triticum*, but it is not useful to do this in operational taxonomy within the subtribe because to be consistent it would be necessary to associate, or lump, not only *Aegilops* with *Triticum* but also *Agropyron*, *Secale*, and *Haynaldia* on the basis of hybridization performance, and a recent successful cross has been made between *Hordeum* and *Triticum*. The existing genera of the tribe are well known, long established, and are operationally useful. They should stand.

As wheat growing has become more intensive the species which are grown become fewer in number. Two species dominate the commercial world: *T. aestivum* and *T. durum*. Rarely, the other species might be found as a crop, but often they are observed as weeds in the fields near crops grown in the centre of origin of the genus. While the significance of the species as crops has diminished their importance as sources of desirable genetic material, such as resistance to disease, cannot be emphasized too strongly. For example, Percival states that 'Upon *T. aegilopoides* (*T. boeticum*) and *T. monococcum* the rust is never found. . . .'

In the course of genome analysis of the tetraploid and hexaploid wheats it was found that separate groups could be distinguished within each. Amongst the tetraploids, a wild species named *T. araraticum* Jakuby and the cultivated *T. timopheevi* did not give the same hybridization patterns as the other tetraploids. It was considered that these two species had some genome other than B in their make-up; and it was designated G. The same studies that suggested that the B genome had come from *T. urartu* also led to the conclusion that the second genome in *T. araraticum* and *T. timopheevi* was likewise from *T. urartu* but had in the course of time differentiated sufficiently to become distinct from the genome B of the other tetraploids. In order to indicate this the letters are now superscripted and the Emmer types are $A^eA^eB^eB^e$, whilst the second group, the *timopheevi* types, are $A^tA^tB^tB^t$.

A rare hexaploid *T. zhukovskyi* Men. and Er., more like the *timopheevi* types, has the genomic constitution $A^zA^zA^tA^tB^tB^t$ where the genome A^z has changed sufficiently from the original A genome to be distinct in its own right and different from both A^e and A^t.

The possibilities for the exchange of genetic material between related species and genera are enormous and there is a remarkably large genetic base from which the plant breeder can draw.

The genus *Hordeum*

The barleys, like the wheats, are considered to have a centre of origin within the 'fertile crescent'. It may have been that barley was domesticated even before wheat but this is something for which we have at present no evidence. There are many species within the genus, and in Great Britain there are three native barleys with a fourth introduction being moderately common. All members of the genus are recognized by possessing at each flowering node on the rachis three spikelets—a median, and two laterals. The median spikelet has one complete flower and is fertile, while the lateral spikelets may or may not have a single complete flower. In our native barleys the lateral flowers are at most staminate but are regularly barren. When the median spikelets are the only ones which are complete the plant is called a two-rowed barley, to distinguish it from those in which all spikelets are fertile, giving six-rowed forms. Barley crops are made up of both types; indeed in some instances both two- and six-rowed forms are found in the same field, but this mixing of forms is associated with less advanced agricultures.

The glumes of *Hordeum* are fine and delicate with a long tip which could be considered an awn. Often fine hairs are found along the edges of the glume. Both median and lateral spikelets have glumes, and in the case of extreme reduction of the laterals the glume is the only spikelet structure that is retained. The lemma is broad and slightly inflated at the base, but it is the very long terminal awn which gives this structure its character and confers upon the barley head its almost unique appearance. The palea is boat-shaped and usually it and the lemma invest the kernel closely and in many the ripe grain consists of the caryopsis with the lemma and palea fused to it. However, naked barleys, in which the caryopsis falls free on threshing and becomes the grain, do exist. In other respects the flower is normal.

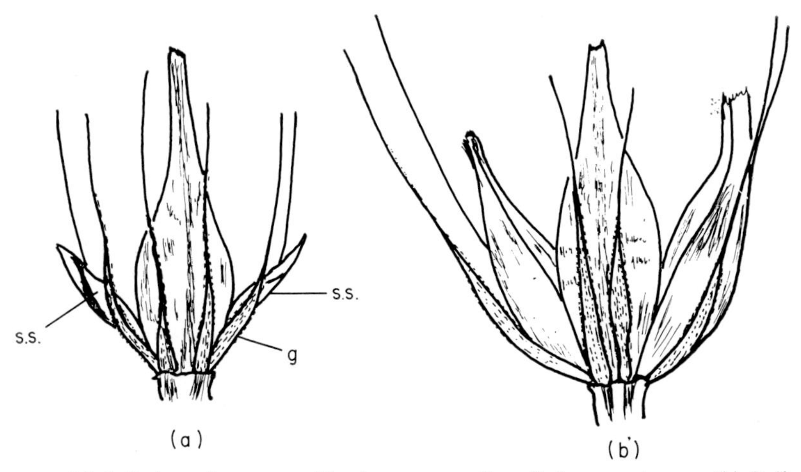

Fig. 11. (a) Spikelets of two-rowed barley: s.s., sterile spikelets; g, glumes. (b) Spikelets of six-rowed barley. Both (a) and (b) at a spikelet node.

As in wheat there are winter and spring forms, and the same considerations apply relating to yield. In barley the contribution of the photosynthetic activity of the head to the grain is greater than in wheat, and it may be that as much as 70% of the carbohydrate comes from this region. This means that small barleys can be as productive as a tall barley or even more so. In the United Kingdom barley breeders have directed their attention towards reducing plant height without reducing yield in order to produce a plant that would not be subject to lodging, that is falling over when nearly ripe into a horizontal position, which could result in difficulties in harvesting.

Two- and six-rowed types (Fig. 11) have already been mentioned as forms of barley, and within the cultivated barleys these forms have been used to designate the main species so that we have

1. *Hordeum distichon* L.—two-rowed barley
2. *Hordeum vulgare* L. emend—six-rowed barley

The emendation of *H. vulgare* from the original Linnaean description is to include both the dense- and lax-eared forms in one species. In the dense-eared types the internode between flowering nodes on the rachis is short resulting in the lateral spikelets of alternating nodes being separate. When this internode is long the head is lax and the lateral spikelets of alternating nodes overlap to give a head which used to be called four-rowed. There is no justification for this separation. The two species of barley recognized by Linnaeus interbreed freely.

Within each of these species there is a number of morphologically distinct forms and these have been given the status of varieties. Whether these should be taxonomically designated variety or forma is a moot point, but the natural relationship is indicated by the fact that there is easy exchange of genes between recognizably distinct forms. The series that can be erected are

H. vulgare
var. *nudum*—naked six-rowed
trifurcatum—hooded six-rowed
inerme—awnless six-rowed
hexastichon

H. distichon
var. *nudum*—naked two-rowed
trifurcatum—hooded two-rowed
inerme—awnless two-rowed
zeocrithon—fan-tail barley
deficiens

In the varieties *nudum* the caryopsis falls free, in *trifurcatum* the long awn is replaced by an inverted sterile floret which forms a hood, shaped like a *fleur de lys* (sometimes these forms are called Himalayan barley), and in *deficiens* the lateral spikelets are gone except for the glumes. In the *distichon* types other than *deficiens* the lateral spikelets are represented by sterile, club-shaped structures and the glumes.

While there are many forms of barley it is found that they are all diploid and it is not possible to employ the same cytological techniques to elucidate the pathway involved in the evolution of this cereal. Two wild species of barley, found in the Middle East, are considered as possible ancestors of today's crop species. *Hordeum spontaneum* C Koch. is two-rowed, and this species occurs throughout the region, being found in the boundary regions of fields. The corresponding wild six-rowed type is *H. agriocrithum* Aberg. but this is not thought to have played a major part in the development of *H. vulgare.* As with the wild wheats these barleys exhibit the brittle rachis, so giving a more fibrous product on threshing, and this has a lower nutritional value. In all probability the modern barleys are no more than a selection from *H. spontaneum* and by the accumulation of sufficient gene differences it has become genetically isolated from the ancestral species.

The genus *Secale*

This is the other main cereal belonging to the Hordeae. At one time it was widely grown especially in the wetter and colder parts of Europe. Not only can it thrive better than wheat in these climates but also it is able to grow better in impoverished soils. As agricultural systems have improved rye has lost ground to the other cereals. However, it forms intergeneric hybrids quite easily with *Triticum*, the hybrid being called *Triticale*, and there is the possibility that this hybrid can be developed as a new crop type extending the acreage on which cereals can flourish.

Rye flour cannot produce a white leavened bread but the dough that is made from it can be baked to give a nutritious loaf, and the grain is acceptable cattle food.

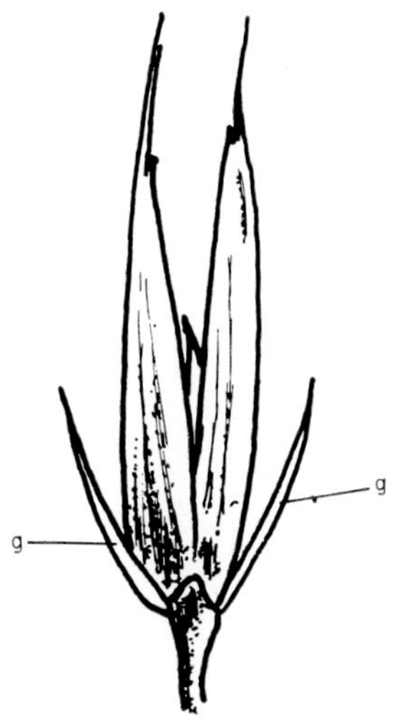

Fig. 12. Spikelet of rye: g, glumes.

Rye has a life cycle like wheat and barley but the winter forms are more common. Some varieties can persist for more than one year but these are more of curiosity value than possible contributors to improving the species as a crop plant.

Rye tillers prolifically and may be used for the production of green fodder. Because of this tendency to abundant tillering the ripening of rye is less uniform than with the other cereals. At flowering the head is a typical spike and the spikelets, arranged opposite each other, may have up to five flowers (Fig. 12). The glumes are narrow, more like those of barley than wheat, and the lemma is awned. The palea is boat-shaped and less tough than that of wheat. At maturity the caryopsis falls free (the spelt character is not known in rye) and the grain is longer and narrower than wheat and is usually a buff green colour. An interesting feature of rye is that the stigma is exerted and remains receptive for a considerable period before pollination occurs. This leads to cross-pollination and also the greater possibility of infection by spores of the fungus *Claviceps purpurea* (Fr.) Tul. If this infection occurs the grain is replaced by the overwintering structure of the fungus—the ergot. Ergots are poisonous and ergoted grain has been responsible for many unpleasant incidents in the history of man. Fortunately this is now rare.

There is only one species of *Secale* grown for its grain, *S. cereale* L. Like barley the origin of the diploid rye is obscure and it is thought that the crop plant is no more than a selection from some wild progenitor. The most likely wild species that could be a possible parental type is *Secale montanum* Guss. though *S. anatolica* Boiss. and *S. segetale* Zuk. have been considered by some to be just as likely to have been the progenitors of rye.

The tribe Aveneae

Amongst the temperate cereals oats have an important place. After wheat and barley it is probably the most widely grown cereal of the cooler parts of the world and it might well move from being of less significance to becoming as widely grown as barley. Plant breeders have not been as successful in selecting high-yielding varieties of oat as they have with wheat and barley. As an animal feeding stuff oat grains are more nutritious than the cereals so far discussed and if the crop could be developed to produce yields approaching those of barley and wheat then there would be a substantial increase in the acreage grown.

Oats belong to the genus *Avena* in the tribe Aveneae and species are widely distributed throughout the world. Some 27 species are recognized and all but one are annual. All are similar, the specific characters being associated mainly with the head and the spikelets (Fig. 13). The crop type is assigned to the species *A. sativa* L. and this is not found in the wild. The description of the genus which follows is based on this species.

In the vegetative state oats follow the course of development associated with an annual grass, either overwintering if a winter annual, or germinating in the spring to produce a plant which completes a growth cycle within six months. The grain of oats consists of the caryopsis tightly enclosed by the lemma and palea, though there are naked species in which the caryopsis is free like bread

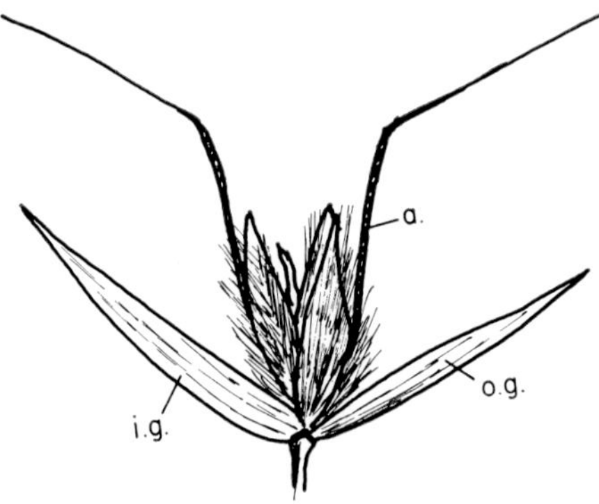

Fig. 13. Spikelet of *Avena ludoviciana*: o.g., outer glume; i.g., inner glume; a, awn.

wheat. The horny lemma and palea constitute the husk while the caryopsis is the kernel, and it is desirable to have as little husk as possible because too high a proportion of husk in the grain reduces its feeding value. Each grain has a pronounced remnant of the rachilla, but the scar on the top of the rachilla varies in shape according to whether the grain in question has been derived from the basal flower or one more distal on the flowering axis. At the base of the grain there is a complementary scar. These scars are most valuable in classifying the genus. The lemma has a heavy dorsal awn, though in the most popular varieties this structure has been selected against, and in any case after the head is threshed the awn is removed so that in commercial samples it is seldom seen. The possession of this dorsal awn which is spirally thickened in its basal portion above which it is usually sharply angled or kneed is a consistent feature in the tribe. This awn is concerned with 'seed' dispersal and burial in the soil.

After germination the radicle erupts through the husks but the coleoptile, enclosing the plumule, grows within them to protrude at the distal end of the grain. In *Avena*, the mesocotylary internode extends almost at the same time as the coleoptile is growing, with the result that the plumular bud is carried away from the grain at an early stage. In this way oat differs from wheat and barley and this slight difference in morphology of the very young seedling, up until about the time when the second leaf is unfolding, is sufficient to provide the basis for using selective herbicides to control weed species of the tribe within crops of wheat and barley.

In the vegetative stage oats are easily identified from the other cereals. The seedling of common oat is without hairs, glabrous, does not have auricles, and has a large membranous ligule. A minor difference between this cereal and the others is the way in which the leaf is rolled. Oat leaves are rolled anticlockwise whereas the others have leaves which are rolled in a clockwise direction.

Oat is the most prolific of the cereals in the production of tillers but, as remarked, this is suppressed in field crop conditions. Each tiller that is produced eventually has a flowering head in the form of a panicle which may be symmetrical and open or clustered to one side. If the latter, the head is said to be secund. Both open and secund heads can occur within a species but it is often used as a good species character.

The spikelets vary in flower number, up to five being found, but in common oat there are usually three. There are also numerous examples where two flowers, one being imperfect, either the proximal as in *Arrhenathrum*, or the distal as in *Holcus*, occur. The terminal flower aborts or produces a very small grain with a high proportion of husk. In every case the flowers are totally enclosed within large papery glumes.

At maturity there are many types of disarticulation, and in wild species the brittleness of the rachilla is marked, but in common oat the tough rachilla means that the grains are retained after harvest. The brittle rachilla of wild oat results in the shedding of grains as soon as the ripe culm is struck leading to ready infestation of fields. Wild oats are now the most troublesome weed of the temperate cereals and this morphological feature of dispersal, along with a complex dormancy, results in an ideal combination for success as a weed.

The origin of oat

The cultivated oat, *Avena sativa* L. is not found wild and has been shown to be hexaploid, with $2n = 42$. Diploid species with $2n = 14$ and tetraploids with $2n = 28$ are known and there is evidence that some of them were cultivated in early times. Genomic analysis like that carried out in *Triticum* has been done and the hexaploid has the constitution AACCDD which is the same as *A. sterilis* L. and *A. fatua* L. Because of the occurrence of rare aberrant types in *A. sativa* with grain features like *A. fatua*, the so-called fatuoids, many considered that common oat was no more than a selection from *A. fatua*. Evidence obtained mainly from hybridization studies points not to *A. fatua* but to *A. sterilis* as the ancestor which gave rise to *A. sativa*.

Until fairly recently *A. strigosa* Schreb. and *A. brevis* Roth. were grown as crops colloquially known as small oats. They are able to withstand more severe climatic conditions and thrive in poorer soils than common oat, but with changes in farming patterns little if any of these are grown. *A. nuda* L., the naked oat, is grown but its yield of grain is low.

While the tribe contains a large number of genera and is well represented in the grasslands of the world only *Avena* is of commercial importance. Others which might be thought of as possible grasses for forage uses are of little consequence, and a few of them are pernicious weeds. As well as the wild oats mentioned there are *Alopecurus*, *Arrhenathrum*, *Holcus* and *Deschampsia*.

The tribe Festuceae

To this tribe belong most of the important forage grasses of the temperate regions of the world. Towards the warmer parts, other grasses belonging to other tribes become prominent, and the Festuceae are displaced as the dominant members of the grasslands.

The salient features of this tribe are the many-flowered spikelets carried in panicles, but in some genera the head is contracted to a spike-like panicle. In the genus *Lolium* the head is contracted to such an extreme degree that it forms a pseudo-spike. In early classifications the *Lolium* head was not considered different from a true spike and this genus was placed in the Hordeae. The pseudo-spike differs from the true spike in that the spikelets, while arranged alternately and diagonally opposite each other are inserted on the rachis radially and not tangentially. The lower glume, except in the topmost flower, is adnate to the rachis so that superficially the spikelet appears to be subtended by a single glume. That the genus is in fact a member of the tribe Festuceae is substantiated by its ability to cross easily with certain members of the genus *Festuca*, primarily *F. pratensis* Huds.

The many-flowered spikelets have glumes which are not larger than the spikelet and are horny, not membranous. In this way the tribe differs from the Aveneae. It should be stated that there are genera of the Festuceae with reduced numbers of flowers per spikelet but these can be identified by the form of the glumes.

The flowers are normal except in the genus *Vulpia* where the number of stamens may be reduced to one, but this genus is so close to *Festuca* in the rest of its morphology that there is no difficulty in assigning it to this tribe.

Within this tribe, indeed within a single genus, both annuals and perennials exist, and of the perennials both bunch and creeping grasses are found. In the composition of a grassland it is observed that when there is sufficient moisture to support a dense vegetation both these growth forms occur and there is much more complete cover of the soil than when the grassland is found in drier regions of the world. In these dry parts the creeping grasses are not well represented and the bunch grasses constitute the grass component of the community. So much is this the case that these grasslands are called the bunch grasslands. The annual grasses do not play a role in the establishment of the world's grasslands, but as pioneers they are very important in the establishment of plant communities, and in agriculture this characteristic makes many of them weedy, though a few of the annuals have been selected as short-term fodder crops, as have the biennials. These short-lived types often are more responsive to applications of fertilizer and so give higher yields than the perennials.

The life cycle of the festuca type grasses follows the normal pattern. The seedling tillers profusely and in the case of biennials and perennials does not flower in its first year of growth. As a result of this vigorous vegetative growth a sward is produced. Swards are consolidated if the grass is grazed or cut for this result in the production of more tillers and because of the activity of the intercalary meristem the plant is not depauperated by the loss of leaf material.

Hairiness of the members of the Festuceae, where it occurs, is sparse and most of the genera are quite palatable. This means that in different climatic and soil conditions one of the genera will be found to be more suitable than the others. In the United Kingdom it is accepted that *Lolium* spp. are the best pasture grasses, but this is because British agriculture is geared to using high levels of nitrogen. Where less nitrogen is employed and the ley system of farming is not practised one of the other genera may prove more valuable, as is the case in certain parts of the Eastern United States of America where *Poa* is the most important genus.

The genus *Lolium*

Within the rye-grasses there are annual, biennial, and perennial species. These grasses are not related to rye and their present name is a corruption of the older name ray-grass. All members of the genus *Lolium* are glabrous, and most of them are quite palatable; only after the production of a flowering head is there a reduction in digestibility. The annual darnel, *L. temulentum* L., the tare of the

Bible, is poisonous when it has ripe grain but its toxicity may be due to a fungus which develops extensively within the grain.

In any species there is considerable variation amongst the seedlings, some being erect and others being prostrate. Even with rigorous selection this variation is present, and in this respect the forage grasses differ from the cereals. The reason for this variation amongst seedlings is due to the outbreeding nature of rye-grass populations. With selection and isolation a degree of uniformity can be fixed in a sample so that the morphology of the seedlings is alike and the sample can be considered as a race or strain, and though now given cultivar status, the cultivar description applies to a phenotype and not a specific genotype.

The erect phenotypes are found amongst those selected for hay, and along with the erect habit there are associated long internodes, reduced leaf production and usually early flowering. Prostrate forms are found most often in pastures subject to intensive grazing over a long period, and we find that these forms have short internodes, are leafy, and flower relatively late in the season. It is also noted that the erect hay types are not so long-lived as the prostrate pasture forms.

Young plants of rye-grass have shiny adaxial surfaces to their leaf blades, and while not unique this is valuable in identifying them. There are small auricles and a small ligule where this blade meets the sheath.

At flowering the head forms the pseudo-spike and the individual spikelets have from four flowers upwards to fifteen. The lower glume is adnate to the rachis and forms a characteristic depression, except for the terminal spikelet where it is free. The lemma may be awned or, as in the important *L. perenne* L., awnless, and the lemma has a round back making it characteristically boat-shaped. The palea is fine and is enclosed by the incurved margins of the lemma. There are three stamens, two lodicules and the single carpel. The grain is the caryopsis enclosed within the lemma and palea. There is a distinct rachilla.

Other Festucoid grasses which are deliberately grown for forage, lawns or soil binding include the following:

Agrostis canina L. subsp. *canina* Velvet bent
A. canina L. subsp. *montana* Hartm. Brown bent
A. gigantea Roth. Red top
A. stolonifera L. Creeping bent
A. tenuis Sibth. Brown top bent
Alopecurus pratensis L. Meadow foxtail
Ammophila arenaria (L.) Link Marram grass
Bromus catharticus Vahl. Schrader's Brome, rescue grass
B. inermis Leyss Hungarian Brome, smooth Brome
Cynosurus cristatus L. Crested dog's-tail
Dactylis glomerata L. Cocksfoot, Orchard grass
Elymus arenarius L. Sea lyme grass

Festuca arundinacea Schrad. Tall fescue
F. pratensis Huds. Meadow fescue
F. ovina L. Sheep's fescue
F. rubra L. subsp. *rubra* Red fescue
F. rubra L. subsp. *commutata* Gaud. Chewing's fescue
Lolium multiflorum Lam. Italian ryegrass
L. rigidum Gaud. Wimmera ryegrass (this species is perhaps a hybrid between *L. perenne* × *L. multiflorum*
Phalaris arundinacea L. Reed canary grass
Ph. canariensis L. Canary grass
Ph. tuberosa L. Toowoomba canary grass
Phleum nodosum L. Small cat's-tail, lesser Timothy
P. pratense L. Cat's-tail, Timothy
Poa compressa L. Canada blue grass
Poa pratensis L. Smooth-stalked meadow grass (S.S.M.G.), Kentucky blue grass
Poa trivialis L. Rough-stalked meadow grass (R.S.M.G.)

The Subfamily Panicoideae

The grasses which belong to this taxon are widely distributed in the tropics and subtropics, but scarcely represented in the cooler regions. This geographical distribution was at one time considered sufficiently diagnostic to be included in the features that were attributes of the panicoid grass. We would expect that amongst the tropical crops the grasses that are utilized would, in the main, belong to this subfamily; this is so. Grasslands as known in the temperate regions do not occur in the tropics. Mostly, the tropical grassland is made up of high grasses and in the grassland there are trees and shrubs, the number being determined by the amount of available water. The more rain the greater the number of woody species and the fewer the number of grasses. It is considered that grass may have evolved within the moist shady forest and that as it was subject to selection it was able to establish and colonize the drier lands to the north and south of the tropical belt. Again we note that the grass is a plant of a community in which the amount of available water is insufficient to support tree growth.

The features that distinguish the panicoid grass have already been discussed, but for the success of the plant in its natural habitat certain aspects of its morphology should be noted. Unlike the festucoid grass the panicoid does develop a small stem in its early juvenile vegetative stage. While this makes it more susceptible to damage from the grazing animal, the growing point is still protected if the tendency for the animal is to browse the leaves. These leaves are relatively broader than the festucoid leaf and they are held on the plant in such a way that they sweep outwards from the culm in an approximately semi-circular fashion. The distichous arrangement of the leaves means that the blades form a canopy which begins to approach a hemisphere. This contrasts with the more upright blades of the festucoid grass, but when the average azimuth of the sun in tropical zones is considered it is seen that the leaf arrangement just described results in the optimum interception of solar radiation.

On the whole panicoid grasses are large and the culms are robust, with solid internodes. There would be a natural tendency for these stems to be bowed down by their own weight or laid under the stress of winds and rains. The panicoid grass can develop from the basal nodes just above ground level a set of specialized adventitious roots which help support the culm. These are called *strut roots*. Internally these strut roots are heavily thickened and they seem not to have any absorptive functions. In maize they are particularly well developed.

The six tribes listed by Bentham as members of the Panicoideae are now considered invalid. Some have been upgraded to subfamily level and in the remainder there has been a degree of consolidation. Of the original tribes only the Paniceae remains as described and the Maydeae are now included in the Andropogoneae. We are then left with but two tribes in this subfamily—the Paniceae and the Andropogoneae.

The tribe Paniceae

In the Paniceae the glumes are typically soft. The individual spikelets are either two-flowered with the lower flower barren, or staminate, the upper perfect or reduced to a single flower. In many there is only the upper glume which encloses the whole spikelet and this is soft and membranous. The lemma of the lower flower is of similar texture but that of the upper, fertile flower is hard and shiny. Sometimes the spikelets occur in pairs and are distinctly pedicellate, the lower being shortly stalked (Fig. 14). A few genera produce bristles from the base of the flowering axis and these confer upon the head a distinct character, e.g. *Setaria italica* (L.) Beauv. (Foxtail millet).

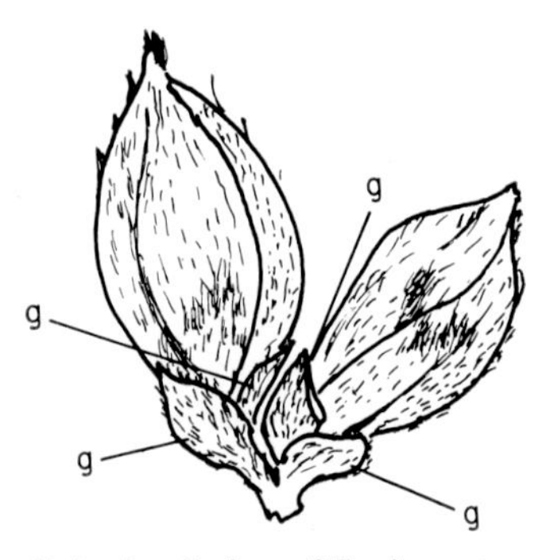

Fig. 14. Paired spikelets of *Panicum sp*: g, glumes.

Species which at one time were important for grain production but are now of only local consequence, are *Echinochloa frumentacea* (Roxb.) Link (Japanese barnyard millet), *Panicum miliaceum* L. (common millet), *Paspalum scrobiculatum* L. (Kodo millet), *Pennisetum typhoides* (Burm. f.) Stapf. and Hubbard (bulrush millet) and *Setaria italica* (L.) Beauv. (foxtail or Italian millet).

However, in the exploitation of the tropical and subtropical grasslands we note that amongst the main contributors to the productivity of these communities there are the following species:

Axonopus compressus (Swartz) Beauv. (savanna or carpet grass)
Brachiaria brizantha (Hochst.) Stapf. (palisade grass)
Br. decumbens Stapf. (Surinam grass)
Br. mutica (Forsk.) Stapf. (Para grass)
Melinis minutiflora Beauv. (molasses grass)
Digitaria decumbens Stent. (Pangola grass)
Panicum maximum Jacq. (Guinea grass)
Paspalum conjugatum Berg. (buffalo grass)
Ps. dilatatum Poir. (Dallis grass)
Ps. notatum Fleugge (Bahia grass)
Pennisetum clandestinum Hochst. ex Chiov. (Kikuyu grass)
Pn. purpureum Schum. (elephant grass)
Setaria sphacelata (Schum.) Stapf. and Hubbard (golden Timothy grass)

Other species from these and related genera are noxious weeds.

The tribe Andropogoneae

These are usually large grasses with solid internodes, annual or perennial. Some of the perennials are creeping. A few genera are of smaller plants.

The flowering heads are paniculate with the spikelets arranged in pairs, one sessile, the other pedicillate on the flowering axis. The glumes are hard, the lowermost one large enclosing the whole spikelet, whilst the lemma and palea, when present, are membranous, but often these structures are much reduced or absent. The spikelets may have one or two flowers and when the latter, the lower flower is staminate or sterile. Within the tribe there are species in which it is observed that the distribution of staminate or carpellate flowers can be within a spikelet, one or other of the paired spikelets, within a flowering head, or in separate inflorescences. The grain is normally shed as a caryopsis enclosed within the glumes since disarticulation occurs below these structures. With modern threshing machinery it is possible to have the caryopsis without its enclosing structures, making the grain of commerce.

The solid stems can act as storage organs and in the genus *Saccharum* substantial amounts of sucrose are retained in the internodes prior to the flowering. If harvested at this time the sap can be extracted and processed to obtain sugar.

Members of the genus *Cymbopogon* produce as secondary metabolites the lemon grass oils whilst in *Vetiveria* another essential oil, oil of vetiver, is obtained by distillation of the roots. Both these products are important in perfumery.

Some species of this tribe are valuable forage plants of the tropics but undoubtedly the great contributors to agriculture from this tribe are maize, sugar cane, and sorghum.

The genus *Zea*

Maize originated in the New World and was not known in Europe until after the 15th century voyages of discovery. In 400 years it has spread worldwide, until it is now the third most important grain crop, and each year its geographical range, for grain production, increases, Even in Great Britain it is now considered to be profitable crop in the warmer south-east of the country, and beyond the area in which it can be grown for grain this plant can still be thought of for the production of silage. Where it can be grown it is highly productive, as are a great many of the other species in the Panicoideae, and it is considered that this potential for high levels of production is associated with the type of photosynthesis carried out by these plants: the C_4 mechanism of carbon fixation.

Maize is a tall, vigorously growing annual, which fairly soon after seedling establishment begins to produce a culm. The young seedling has a characteristically spathulate first leaf; succeeding leaves are relatively broad with a well-defined central region. The leaf may be glabrous or sparsely hairy though frequently hairs are prominent along the margins of the upper region of the blade. The blade can exceed 1 mm in length, and such large leaf blades can be damaged severely by hail. In fact hail can be the major hazard to maize production in certain parts because loss of leaf material prior to development of the grain usually results in the death of the plant.

Maize is planted as a row crop with seeds sown individually and widely spaced. Such circumstances in other grasses would lead to abundant tillering but with maize it is unusual to find more than two tillers per plant, so that at any one station, or hill, we expect to find three stems.

At flowering the plant produces two types of flowering head. A terminal panicle, similar to that produced by many grasses, is formed but the spikelets carry male flowers. This terminal head is called the tassel. The other flowering head is produced in the axils of those leaves in the mid-section of the culm and these produce female flowers. The axillary heads are called cobs, and because of their structure (q.v.) the stigmas are long and must extend beyond the bracts covering the female flowers. These stigmas are called *silks*. Amongst grasses this monoecious habit is almost unique and in maize it has led to a very important cultural improvement, the use of hybrid corn.

All plant and animal breeders know that hybrids are usually more vigorous than their parents. This vigour is called *hybrid vigour* or *heterosis*. Conversely, continued inbreeding leads to a loss of vigour. Under natural conditions maize would be outbreeding to a certain degree but when brought into intensive cultivation there would be a tendency for pollination to take place amongst the members of the same stock, or even to have self-pollination as the rule. This circumstance prevails in many plants which do not have a mechanism to prevent or avoid self-fertilization, but with small hermaphrodite flowers there is little that the breeder, or producer of seed can do to maintain directed outbreeding. With maize the situation is different. Pollen can be either collected from one plant to be used in pollinating another, or pollen can be excluded from the silks so that directed hybridization can take place. A simple method has been developed to produce hybrid material. Plants of the parental types are sown in alternating strips in the field, and the tassels of those plants to be used as female parents are removed before anthesis, while the tassels on the male parents are allowed to ripen. When the pollen is shed on to the receptive silks of the female parent cross-pollination is effected. The only difficulty is to ensure that the pollen of the male parent is ripe at the same time as the stigmas of the female parent are receptive.

The staminate flowers of the tassel are carried in pairs in each spikelet, and the spikelets themselves are arranged in groups of two. One of the spikelets is sessile on the flowering axis whilst the other is carried on a pedicel, so reflecting the common arrangement of spikelets in the Panicoideae. In maize the glumes are robust and the lemmas and pales of each flower, though well developed, are delicate and quite membranous.

Contrasting with the staminate flowers in recognized spikelets, the pistillate flowers of maize are sessile on a robust flowering axis which at first glance appears to have the flowers arranged in distinct longitudinal rows numbering from eight to twelve. On closer examination it is seen that the rows are, in fact, paired and that this pairing is a reflection of the paired arrangement of the pistillate spikelets. Each spikelet has two flowers which are subtended by short glumes, which in the older heads are not readily seen. The lower flower is sterile and the upper flower possesses a functional ovary. Both flowers have short broad membranous lemmas and paleas, with perhaps rudiments of the stamens occurring. In the fertile floret there is no evidence of lodicules, but the really distinctive feature is the very large development of the stigma which grows out from the top of the ovary and in some instances can be 50 cm long.

The young cob is well protected by a number of modified leaves arising from the basal nodes of the flowering axis. The modified leaf can be considered a bract and these bracts persist throughout the period when the grains are maturing and at maturity wither and remain tightly enclosing the ripe head, even after its removal from the mother plant. The husks, as they are called, and the stalk, or central portion of the cob, constitute a bulky residue which is lower in feed value than the grains themselves, and are not normally fed to cattle. If maize is harvested for silage it can be allowed to reach a stage when the grains are filling and though there are husks and stalk the ensilage process renders them acceptable to animals.

Maize, as remarked, was not known in Europe until the 15th century, and in

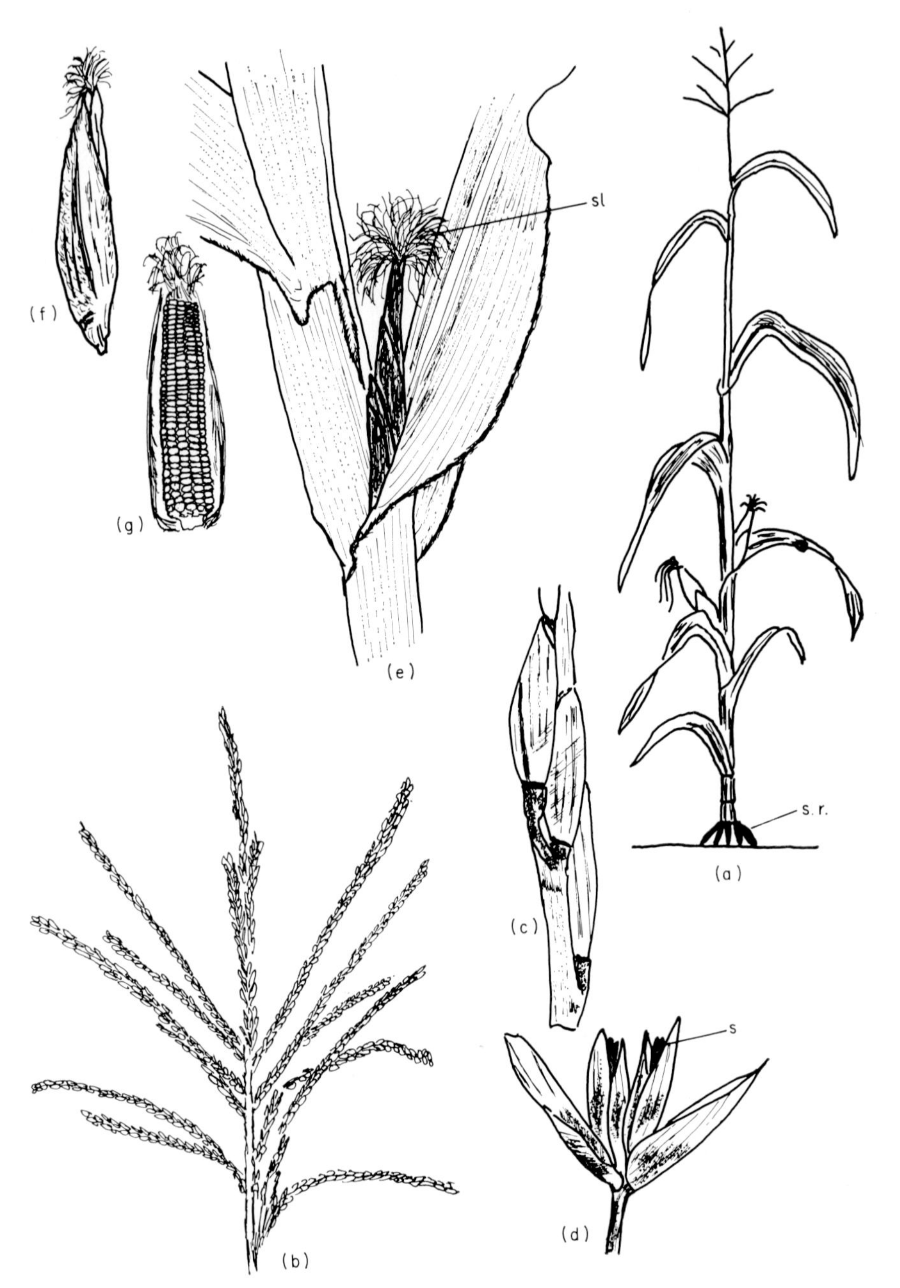

Fig. 15. Maize. (a) Adult plant: s.r., strut root. (b) Tassel. (c) Arrangement of paired male spikelets on tassel rachis. (d) Stalked male spikelet opened to show two flowers: s, stamens. (e) Cob in leaf axil: sl, silks. (f) Whole mature cob with sheathing bracts. (g) Bracts removed to show regular rows of grains.

the New World all the evidence available indicates that it does not occur as a wild plant. The plant was called *Zea mays* by Linnaeus though the generic name *Zea* had been used for certain wheats. It has been generally accepted that *Zea* is monospecific but certain authors now include an annual grass of Mexico in this genus (*Zea mexicana* Reeves and Mangels). This annual grass is called teosinte and it occurs wild though it is sometimes utilized by the peasants. Its importance to us lies in the fact that hybrids between maize and teosinte are not infrequent, and *Zea* can also hybridize with another grass (gama grass, *Tripsacum dactyloides* L.—a perennial weed, also of tropical and subtropical North America). It is perhaps better to consider teosinte as belonging to the genus *Euchlena* and to be designated *Euchlena mexicana* Schrad., and to keep *Zea* monospecific. We have already seen how intergeneric hybridization is common in the grasses and if all cases where hybrids can be produced are used to extend a genus and remove other good genera grass systematics would be difficult indeed because it would be difficult for the agrostologist to be sure of including all the necessary information in describing a genus.

Teosinte has the same habit as maize with a tassel and axillary cobs, though everything about the plant is diminutive compared to maize. The pistillate heads are individually small but more numerous, and the glumes of the spikelets are large. This same large glumed character is found in some maize mutants, and the oldest maize remains have small cobs with large glumes, in other words very like teosinte. We shall never know directly if maize is no more than a selection from teosinte, in which case the inclusion of teosinte in *Zea* would be justified, or whether, as has been suggested, that teosinte is a derivative of maize! In this case inclusion of the two plants in the same genus is likewise justified.

Gama grass is different in that there is no separation of pistillate spikelets into separate axillary cobs. These spikelets occur in the basal part of the head and are sometimes partially buried in the rachis. Maize crosses with this plant and it has been suggested that a hybrid of these two was the progenitor of teosinte.

The vexed question of the origin of maize cannot be left without noting that it is not uncommon to find staminate spikelets in axillary heads, and pistillate ones in the tassel. The unique type of monoecism can break down and these teratological forms might indicate that maize is a derived type, from an unknown ancestor, the present cultivated forms having been selected from a wild maize which is now extinct.

An Old World species of the Andropogoneae which is grown for its grains is *Coix lachryma-jobi* L., Job's Tears or adlay. While a robust, freely tillering annual not having the stature of maize, its flowering head is interesting and has some affinity to that of *Tripsacum*. These flowering heads arise from the axils of the upper leaves and have some ten to twelve spikelets, the lower one or two being enclosed in a hard urnicate structure. The spikelets within these are one-flowered, the flower being pistillate. Beyond the basal spikelets up to ten staminate spikelets, each with two flowers, are produced. The male spikelets fall readily from the head to leave at maturity the grain enclosed within the developed urn-shaped bract which can be variously coloured.

Adlay, while of local importance, is more a curiosity than a crop which enters commerce. The same cannot be said of the next member of the Andropogoneae which we will consider.

The genus *Saccharum*

Sugar cane is grown throughout the warm tropics where the soil is sufficiently fertile and the hydrologic situation provides sufficient moisture without there being waterlogging. Sugar cane belongs to the genus *Saccharum* L. and of the six species which are recognized two are considered to be wild, the other having originated in cultivation. The most widely grown cane is *S. officinarum* L. or noble cane, though in parts of India both *S. barberi* Jesweit and *S. sinense* Roxb. are cultivated.

Saccharum consists of robust perennials with well-developed culms terminating in a paniculate head called the *tassel* or *arrow*. The internodes are solid and within them substantial amounts of sucrose are deposited prior to flowering (the sap containing almost 20%). The leaf sap is not rich in sucrose, most of the sugars present being monosaccharides, so at harvest the leaves are removed either by slashing or burning.

The arrow has a distinct feathery appearance and in this respect is like the inflorescences of *Imperata* spp., *Erianthus* spp., and *Miscanthus* spp., genera with a more widespread distribution than the south-east Asian *Saccharum*. The feathery appearance is due to the production of long silky hairs below the paired spikelets, one of which is pedicellate. The arrangement and structure of the spikelets accords with that described for the tribe. In *S. officinarum* the flowers are generally sterile but seeds can be produced and there is the capacity for this species to cross with others. However, seed is so rare that the crop is established by planting parts of the underground stem.

While the cane does produce seed the method of propagation has remarkable similarities to that employed for the potato of the temperate regions (q.v.) with the result that in practice there has been established a number of clones, and to allow us to appreciate how this cloning is carried out it is necessary to describe the structure of the sugar cane plant.

As remarked, a portion of the underground stem is used to propagate the plant, and this is called a *sett*. A sett consists of a number of short internodes, and corresponding nodes. At each node there are incipient root primordia, an axillary bud and, if from below ground, a scale leaf. Just above the node is an intercalary meristem. If an aerial part of the axis is examined the same arrangement can be seen though the scale leaf is replaced by a true leaf, and in fact setts can be made from aerial portions of the axis, though they do not establish as rapidly as those from below ground regions.

On planting the sett the root primordia are activated and a series of adventitious roots are formed. These establish the plant. The axillary bud grows out and develops as a short rhizome, its subterranean nodes repeating the behaviour of the nodes of the sett. Eventually the terminal buds of the rhizome system become negatively geotropic and grow upwards to give aerial stems. All the short rhizomes do this and subsequent order branching from the first rhizomes repeat the pattern so giving a clump of stems. As with the cereals, it is desirable that the culms constituting a clump should be closely similar in age to aid in harvesting, so planting densities of setts are arranged to do this. On flowering a culm dies and the plant is perennated by there being released an outgrowing of subterranean buds that were not active in the development of the first crop of culms. The first cycle of growth usually occupies more than one calendar year—as much as 18 months in some cases—but the subsequent crop, called the *ratoon* crop, is taken within a calendar year. The cropping is continued from ratoons in succeeding years. (A ratoon is a shoot which develops from an underground stem after the main stems have been cut down.)

The plant is photoperiodically sensitive and the flowering depends on the plant being exposed to short days. The different clones have very precise day-length requirements for flower induction and this feature can be employed in sugar production to limit the freedom with which a clone will flower by growing it in a region where the daylength exceeds that which is necessary for flowering. Since flowering reduces the level of sucrose the prevention of flowering is profitable. Parenthetically it should be noted that the upper portion of the flowering cane has high levels of reducing sugar, and as with the leaves it is removed before the sucrose is extracted. Flowering can be prevented by giving the plants a 'night break', a short exposure to light in the middle of the dark period. This has proved remarkably successful and is now routine on the most progressive plantations.

The noble cane is thought to be a selection from the wild species *S. spontaneum* L. but more elaborate theories regarding the origin of the cultivated cane have been proposed. These include its origin as a hybrid from within *Saccharum* itself or indeed with one of the related genera mentioned earlier.

The genus *Sorghum*

While this genus is now found throughout the warmer regions of the world its centre of origin is probably north-east Africa, though the grain sorghums may have had a secondary area of origin more to the west and south in that continent. As a grain crop sorghum now comes fourth to wheat, rice and maize and the acreage grown is increasing annually as the crop is improved. Taxonomically the genus presents a number of problems no doubt due to the freedom of outbreeding and the production of interspecific hybrids. In subsistence agriculture this breeding system results in crops which are very variable, and in the areas where this is still the practice not only is the variation manifest within a crop but the system of cropping provides many disturbed habitats, which allows the establishment of progenies from hybrids. Perhaps it was this variation that made agriculturalists in areas where commercial agriculture was practised reluctant to take up this genus as a major crop. However, once selection had been undertaken and uniformity introduced the full potential of sorghum was realized.

In the mid-1930s Snowden examined a large collection of African and Asian Sorghums and he recognized 31 cultivated species of which 12 were considered as major grain crops, but we should now look upon Snowden's classification as having been too fine since he accorded specific status to variants that were probably no more than segregants of the outbreeding population. For this reason it is better to restrict the number of species that are considered as grain sorghums, perhaps to the ultimate reduction of one—*Sorghum bicolor* L.

If we accept that the sorghums can be included within a single species we must also recognize that the range of variation is large and that those variants with distinctive morphological features can be assigned either varietal status or be thought of as cultivars. Perhaps the best example of this is seen when a comparison is made between milo with its very compact panicle, and broomcorn with panicles which are so open and with long laterals that the head is in fact employed as a whisk broom.

Sorghum is a typical panicoid grass sometimes with a single main culm, but able to produce some tillers, but tillering is not progressive as with most other species of grass. Strut roots are produced but the plant is not as tall as maize and the modern cultivars are short, though with the employment of hybrids they are usually robust.

The flowering head is a panicle varying, as remarked, from compact to spreading. The spikelets are carried in pairs, but the terminal group of spikelets on any rachilla is made up of three, a not uncommon arrangement in panicoid grasses. The sessile spikelet of each pair is complete and fertile while the pedicellate spikelet is incomplete or sterile.

The sessile spikelet has one lower flower which is sterile but the upper flower contains both stamens and carpel. The glumes are large and have distinct central keels. The two glumes afford considerable protection to the fertile upper floret and their protective function is supplemented by the large lemma of the lower flower, which is the only developed structure of this flower. The lemma of the upper flower is membranous, sometimes awned, and has a split apex. The palea is delicate but may be absent.

The pedicellate spikelet is variable, and may even be *caducous*, that is it falls off without ever developing properly, or it may produce an upper flower which gives rise to stamens that produce functional pollen.

Sorghums are grown as fodder and in some there is present a *cyanogenetic glycoside* called dhurrin (one of the sorghum races is dhurra corn) and it is possible for enough to be ingested to cause digestive upsets, or even death. The hazard only exists if the sorghum is eaten as fresh leaves. (A fuller treatment of *cyanogenesis* is given on p. 83.)

Sorghum species other than *bicolor* are grown as fodder crops but a few introductions, mainly of perennial species have proved to be unsuccessful because the species has escaped and become weedy. This is true of *S. halepense* (L.) Pers., Johnson grass, in the United States. Most species of *Sorghum* are cyanogenetic, but on being made into hay the capacity to produce poison is lost.

Other members of the Andropogoneae that are cropped are some species of *Andropogon*, *Themeda triandra* Forsk., *Hyparrhenia rufa* (Nees) Stapf., *Ischaemum indicum* (Houtt.) Merr., all considered satisfactory pasture grasses in their own region, and *Cymbopogon* spp., and *Vetiveria zizanioides* (L.) Nene producing essential oils, *viz.* lemon grass oils and oil of vetiver respectively.

The Subfamily Oryzoideae

While some authors consider that there are grounds for instituting a number of tribes in the Oryzoideae it is generally accepted that all the genera of this subfamily can be included in a single tribe, the Oryzeae.

The oryzoid grasses are found either as true aquatic plants, or on soils which are well supplied with moisture. The common rice, *Oryza sativa* L. is grown as an emergent aquatic during its seedling and early adult life; only at the ripening stage is there removal of water from the soil. Rice which is partially submerged during part of its life is called paddy rice. Rice can be grown not as an aquatic but on dry soil for its whole life, and this crop is called upland rice.

The main genus is *Oryza*, but in North America *Zizania aquatica* L. was harvested as a grain crop by the Indians though it was never cultivated. Today the collection of the grain of this wild rice is by statute reserved for the Indian population.

The genus *Oryza*

Rice is the second most important cereal economically. While a substantially smaller acreage of the world is devoted to the growing of rice than to wheat, the difference in the amounts of wheat and rice that are produced is not as great. This is because in much of the rice-growing regions it is possible to obtain two crops per annum.

Rice, *Oryza sativa* L., is a variable grass. It is not found wild and some authors have considered that today's cultivated rice has originated more than once; it is *polyphyletic*. This view is not widely held, and it is considered that *O. sativa* along with *O. glaberrima* Steud., the other cultivated rice found mainly in Africa, had a common ancestor in *O. perennis* Moench emend. Sampath. This last species was also a possible ancestor of other wild species, which are found today as weeds in rice fields and which can cross with the cultivated species.

The grain of rice is the caryopsis enclosed in a horny lemma and palea with the remnants of the small glumes attached. As such it is an unattractive foodstuff and the outer coverings are removed to make it more palatable. If only the husks are removed the product that is left is called brown rice, but even this is not accepted in some societies which demand an even more refined material. To produce this from the brown rice, which is the caryopsis, the pericarp and the aleurone, along with most of the embryo, are removed. What is left is polished rice which is almost exclusively carbohydrate and, while a filling food, is it not nutritious. In regions where the population will not accept brown rice, which is a highly nutritious food, polished rice is made, and then the vitamins removed are added back as a solution to coat the grain.

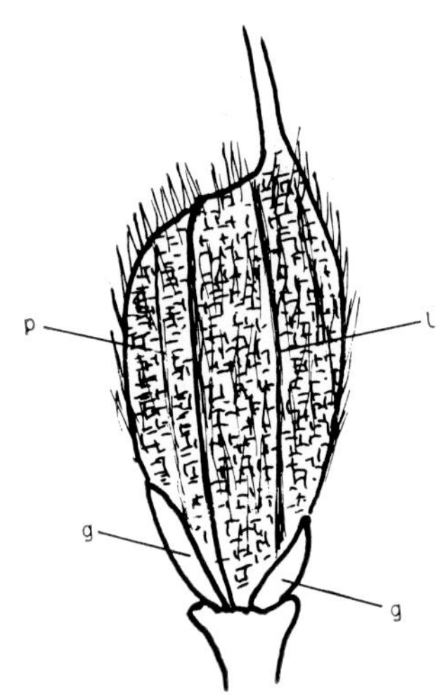

Fig. 16. Spikelet of rice, single-flowered: g, glume; l, lemma with awn; p, palea.

Rice is a free-tillering annual, and since it is spaced on transplanting from the seed bed into the flooded paddy its capacity to tiller is most valuable. Germination can take place under water, but it has been found that planting out in the flooded paddy gives a surer establishment.

The root system of rice is like that of the typical grass, and not similar to the root system of a hydrophyte (helophyte) though at inflorescence production and later, the roots that are produced are superficial and can be diageotropic. It is after these roots are formed that the water is removed. As an aside it may be said that the value of growing rice under water in the tropics is that there are fewer aquatic than terrestrial weeds to contend with and that flooding rice was an early method of selective weed control. Also growth of free-living nitrogen-fixing blue-green algae occurs in the paddies and it has been calculated that these can contribute as much as 20 kg of nitrogen per hectare.

The stem is erect and may have both solid and hollow internodes. The leaves are robust and usually asperulate along the blade margins. Many statures are found within the cultivars from the modern extreme dwarfs with upright spikey leaves to the older very tall varieties with drooping leaves, but in all of them the flag leaf is distinctly different from the other culm leaves.

At flowering rice produces a panicle with many single-flowered spikelets which arise on short stalks. Each spikelet has two small membranous structures at its base and these are considered to be glumes but an alternative suggestion is that they represent the lemmas of aborted spikelets. There may be justification for this view since the small membranous structures are distant from the lemma and there are indications of a pair of ridges just below the lemma. These could be the true glumes. The lemma is pronouncedly boat-shaped with prominent nerves, and is covered with spiculate hairs. Cultivars with small twisted terminal awns are found. The palea, though smaller than the lemma, and invested by it, is similarly keeled, prominently nerved and spiculate. The rice flower agrees with the general description given for the grass flower in that it has two lodicules, but there are six stamens. Not all members of the Oryzeae have six-stamened flowers but it is accepted as a tribal character. There is a single carpel with two feathery stigmata.

The Subfamily Bambusoideae

While there are three subfamilies of the Gramineae still to be considered this one, the Bambusoideae, is the only one of substantial import in providing materials which can be used directly by man. The other two subfamilies, Arundinoideae and Eragrostoideae, do have representatives which are utilized as part of natural plant communities, and perhaps as the world's grasslands come under more intensive management members of the Eragrostoideae in particular, will be treated in the same way as certain festucoid grasses. The bamboos are, however, unique.

At first sight the bamboo does not look like the normal grass, especially to those most familiar with festucoid types. Nevertheless, when comparison is made, not with the vegetative festucoid or panicoid, but with the flowering panicoid, a greater similarity is seen to exist. Many of the panicoids are robust, and above each bud there runs an internodal groove, and the internodes are seen since the sheath is not pronounced. The bamboo, during the whole of its life, has a distinct culm, and the internodes exhibit the groove just mentioned.

The bamboos differ in almost every way from the grasses we have considered so far. Many of them are monocarpic but it takes many years for the plant to become ripe to flower. In the vegetative state the plants produce a well-developed subterranean system of axes, which may be either short-jointed, with small robust internodes, and a sympodial growth habit, or an extensive rhizome system in which the internodes are long and the habit is monopodial. The former are called *pachymorphic* and the latter *leptomorphic*. From the rhizomes aerial shoots are produced to give a clump, and this clumping is characteristic. At first the internodes are short and at the nodes adventitious roots may be produced. As the culm develops the internodes lengthen and the plants become robust with an upright stature, a few being stragglers. In the extreme case of *Dendrocalamus giganteus* Munro the culm can exceed 30 m and be about 30 cm in diameter. The internodes can be solid or hollow.

The leaf while sheathing and possessing a blade has a distinct constriction between these two regions (Fig. 17). It is considered that the constriction is at

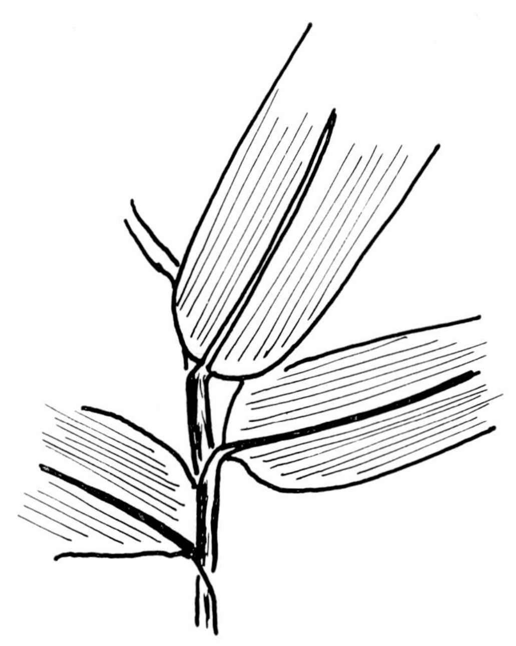

Fig. 17. Basal region of leaf blade of *Arundinaria tecta* (Walt.) Muhl., showing the constriction typical of bamboos.

the base of the blade rather than the top of the sheath, and while it has been interpreted as a petiole it should not be considered as such. In the axil of the leaf there is a bud, but it is the distal buds which develop to give aerial branching and not the basal to give tillers as with other grasses. The secondary branches in turn can give rise to a tertiary system. Eventually, the terminal regions of the branches produce heads. These heads may be open or contracted panicles and the spikelets are one- to many-flowered. Each spikelet has two glumes, and the lemmas of the flower are like the glume, usually without an awn; the palea is frequently with two keels, but may be absent. The protective non-essential, indeed non-floral, parts of the inflorescence are therefore like those of other grasses. When we look at the floral parts of the spikelet differences are considerable, and immediately obvious. There are three lodicules, most often six stamens, though there can be more, but in some species the number of stamens may be reduced to three, and the stamens can have connate filaments. The carpel has a single style from which 1–3 stigmas arise. The fruit can be a caryopsis, but nuts, berries and drupes are found. Those features which might be thought to be closer to the structure given to the classic monocotyledonous flower have led some authors to place the bamboos as the most primitive of the grasses. At most we can say that the bamboo is probably, of the grasses extant, the one most like the progenitor of this great family. To try to deduce more is speculation. There are not many botanists who have seen and examined bamboo flowering heads, so the study of this part of the plant has not been as extensive as could be desired.

As remarked, flowering in bamboos is a rare event and in many of the monocarpic perennials flowering is gregarious, all the plants in a region flowering at the same time. Perhaps because of this shortage of flowers, the 200 or so species in 45 genera are separated mainly on the basis of vegetative characters. Their distribution is principally in east and south-east Asia, with a few in Africa. The most widely distributed common bamboo, *Bambusa vulgaris* Schrad. ex Wendland is not known in the wild, and has spread throughout the tropics as a result of man's activities. It provides, in its woody culm, material for construction. Paper pulp is also a product of this bamboo.

Other species are utilized in the same way as common bamboo but as well as being cultivated, wild growing plants are exploited, and some very specialized products can result, e.g. Malacca canes, split canes for fishing rods, and the young succulent shoots of many species are used as vegetables.

It is difficult to summarize any treatment of the grasses but what may be said of the bamboos might be said of all of them—'If the grasses had not existed they would have to have been invented'.

CHAPTER 2

The Leguminosae: Beans, Peas, Clovers

The Leguminosae is a very natural family of flowering plants. According to some authors it is amongst the largest of the flowering plant families with nearly 700 genera and about 18 000 species but others consider that the family should be split to give three families each based on the subfamilies recognized by most Angiosperm systematists. While there may be good grounds for splitting the Leguminosae into three smaller families there is such a degree of homogeneity within the taxon called the Leguminosae that there is little justification for us to split it in this treatment of the crop species.

The Leguminosae are widespread, having a distribution almost as cosmopolitan as the grasses. Though the members of this family are distributed worldwide we do not meet many plant communities in which the leguminous plant forms the dominant species. We do not encounter terms equivalent to the grasslands. Nonetheless the legume has a very important role to play in the establishment of plant communities, for without them the amount of nitrogen circulating within the biosphere would be much less than it is today.

In their vegetative growth habits the Leguminosae exhibit virtually all the growth forms seen amongst the higher plants. There are small annuals, herbaceous perennials with a variety of overwintering structures, shrubs and trees. Many of the legumes are specialized, amongst them species which are tropical lianas, a type of woody climber with much contorted stems and aberrant secondary thickening. Others possess xeromorphic features which are associated with true xerophytism.

Legume seeds are usually without an endosperm and the testa is well developed. A wide range of size occurs from just over 1 mm to almost 5 cm long and the most frequent shape is that of a bolster with a slight curve. A good many have seeds which are kidney-shaped. The seed is usually, but not always, laterally compressed (Fig. 18). A distinct hilum is seen either near the base or within the arc of the concave side and the micropyle is prominent towards the apex, away from the hilum. The seeds are of many colours, and even within a species varieties can differ in having differently coloured seeds. Some even have seeds which are particoloured.

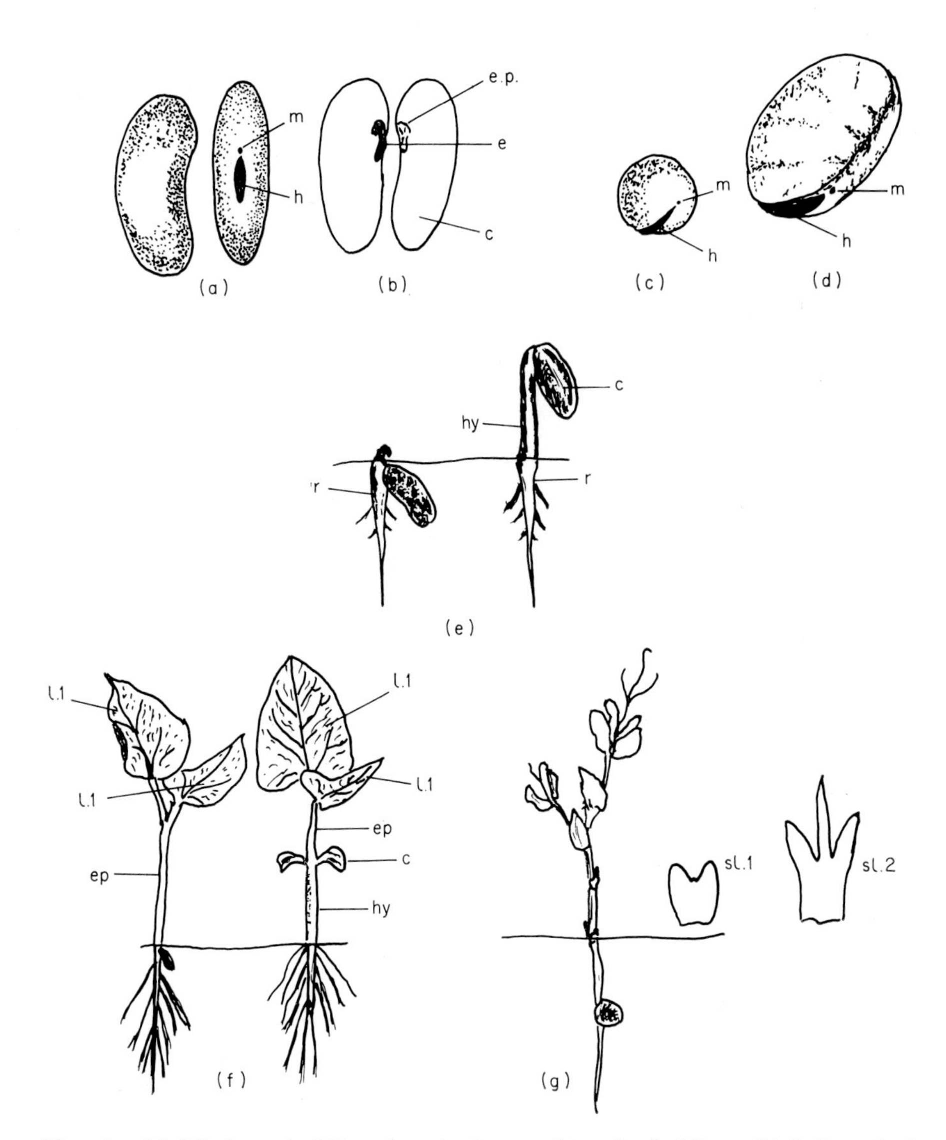

Fig. 18. (a) Whole seed of *Phaseolus vulgaris*: m, micropyle; h, hilum. (b) Split seed of *Ph. vulgaris* with testa removed: c, cotyledon; e, embryo; e.p., embryo pocket. (c) Seed of pea. (d) Seed of broad bean. (e) Germinated seeds of *Ph. coccineus* (on left) showing hypogeal germination and *Ph. vulgaris* (on right) with epigeal germination: hy, hypocotyl; r, radicle. (f) Seedlings of (left) *Ph. coccineus* (right) *Ph. vulgaris*: ep, epicotyl; l.1, paired unifoliate first leaves. (g) Pea seedling: sl 1 and sl 2, first and second scale leaves respectively.

Within the testa there is a well-developed embryo. The two cotyledons act as the food storage tissue and depending on the species the main reserve material can be either starch or oil. The oil content can be high, up to 25%, and the legume seed is a source of fixed vegetable oils. Today one species predominates as an oil crop—soya bean—but others might be developed in the future. On analysis it is found that legume seeds have a high content of protein, on average twice that of non-leguminous seeds with a range of protein concentration from 16.2–47.5%. Legume seeds are therefore most important as sources of protein, and when a legume is grown for its seeds it is called a *pulse crop*. Unfortunately many of the legumes store in their seeds substances which are poisonous, and there is in this family some of the most noxious of plant products to be encountered amongst the flowering plants. The poisonous properties of the legume will be dealt with later (p. 65).

The embryonic axis is small in relation to the cotyledons but is readily discernible. The radicle lies to the outside close to the micropyle while the plumule with its characteristic hook is enclosed within the folded cotyledons. On germination the cotyledons may stay within the testa, and remain below ground to give *hypogeal* germination, or they may enlarge, burst through the testa and with elongation of the hypococtyl be carried above ground to become green and begin to photosynthesize. Germination in this case is said to be *epigeal*. The type of germination adopted is species specific and can be a valuable diagnostic feature. In the seedling the first-formed leaves are unlike those of the adult mature form. In many, the first leaf to be produced has stipules, a petiole, and a simple lamina; this is a *unifoliate* leaf. In others the first leaf is seen to be no more than modified stipules and cannot be considered to be anything other than a scale leaf. The unifoliate leaf usually gives way immediately to the adult form, but in those species which produce a scale leaf there is a progressive change in leaf shape until the final adult form is achieved. However, the nature of the progression, whether the first leaf to be formed is unifoliate or scale, is dependent largely on the form of the adult leaf.

The adult leaf is stipulate, petiolate and compound, but many variants are found. The stipules can be extremely large as in pea or reduced to small, hair-like structures. The petiole is furrowed to a greater or less extent. Both palmately and pinnately compound laminas are encountered in the Leguminosae (Fig. 19) and it is sometimes difficult to assign a trifoliate leaf to its category. Terminal leaflets are often modified to tendrils. The leaves may be *glabrous* or hairy, the latter detracting from its palatability, so reducing the value of the species as a forage crop. In legumes at the bases of the petiole and also the leaflet stalks there are highly modified regions called *pulvini* (sing. *pulvinus*). As a result of turgor changes in these regions the relative positions of the leaf and leaflets with respect to the stem can be changed. These movements occur between day and night and have been called 'sleep' movements but in some species rapid movement is occasioned by shock, e.g. in the sensitive plant (*Mimosa pudica* L.), and by high temperature, e.g. the telegraph plant (*Desmodium gyrans* D.C.). Before going on

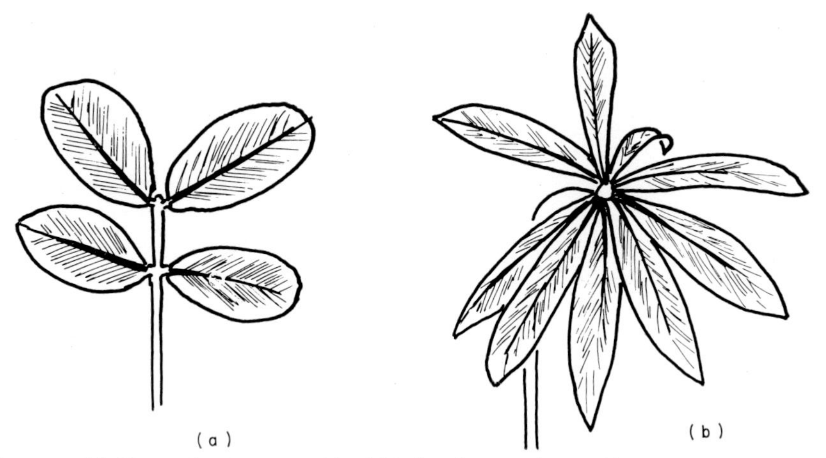

Fig. 19. (a) Pinnately compound leaf blade of groundnut. (b) Palmately compound leaf blade of lupin.

to consider plant form, the final feature to note about the leaf is that, like the seed it has a protein level about twice that of a non-leguminous plant.

With few exceptions, the annual legumes are weak straggling or climbing plants and, as well as employing tendrils which are markedly thigmatropic (Fig. 20), many species exhibit climbing stems. The terminal portion of the stem in these climbers has rapidly elongating internodes and the leaves are not fully expanded. The plants possess an exaggerated circumnutation and when the stem comes in contact with a support, sensory hairs at the point of contact

Fig. 20. Leaf blade of pea with terminal tendrils.

transmit a stimulus which alters the growth pattern of the internode to make it entwine around the support. These tall climbers exhibit pronounced monopodial growth. A species is not exclusively of one habit type for in some the plant may grow sympodially. The internodes in this case are small and while the plant does tend to be lax it is more upright than the climber. The monopodial tall forms, if grown as a crop, require to be staked on poles and such species have come to be called *pole forms* while the sympodial types are known as *bush forms.* The capital outlay in preparing, growing, harvesting and clearing a pole crop is considerable and is only justifiable in horticulture.

The herbaceous biennials and perennials can have a number of growth forms but the most frequent are those where either a rosette is produced which later gives rise to an elongated flowering stem or there is developed a robust woody crown from which there arises in succeeding years crops of adventitious shoots. The basal regions of the previous year's shoots are persistent and axillary buds from them can contribute to the crop of aerial shoots that are produced in any one year.

The herbaceous legumes tend to be deep-rooted and in many species successful growth is dependent on the roots not being impeded in their progress through the soil. In the case of broad bean this is very important and the yield is absolutely related to the length of the primary root. Another species with a deep root system is lucerne and its inability to grow in certain shallow soils is well known.

Like the grasses, some of the legumes have the capacity to produce adventitious roots at the nodes to give creeping perennials, and a few are stoloniferous or rhizomatous.

The woody habit is common, trees and shrubs occurring.

The flowers are most often carried in racemose inflorescences which may be terminal or axillary. Terminal spikes are frequent, as are globose racemes which are called *capitula.* In pea we have an example of a species which has forms with few-flowered axillary inflorescences which are difficult to interpret as racemes but it is generally agreed that the indeterminate inflorescence is the only one found in the Leguminosae.

Cauliflory, where the inflorescences erupt from the stem without being related to a node, is found in the Judas tree (*Cercis siliquastrum* L.).

Within the family a number of flower types are found, but three main groupings can be made:

1. Flowers regular.
2. Flowers zygomorphic, uppermost petal innermost in bud.
3. Flowers markedly zygomorphic, uppermost petal outermost in bud.

Technically we speak of the flowers in group (2) as *imbricate ascending* and in group (3) as *imbricate descending.* The manner in which the calyx and corolla members are arranged in the bud is termed the *aestivation* of the bud, and when the members do not touch the aestivation is *open,* when the members touch it is *valvate*, and in group (1) above, this is the case. Imbricate aestivation occurs when the edges overlap, groups (2) and (3), and ascending indicates that on moving acropetally the succeeding members follow a course inwards on the axis while in the descending series the movement is basipetal with the members following an inward course. This results in the uppermost anterior petal of groups (2) and (3) being respectively to the inside and outside as indicated above.

These features of the flower are used as the basis for establishing the three subfamilies of the Leguminosae. They are: Mimosoideae corresponding to group (1) (also known as Mimosaceae); Caesalpinioideae corresponding to group (2) (also known as Caesalpiniaceae); and Lotoideae corresponding to group (3) (also known as Papilionatae, Papilionoideae, Faboideae and Fabaceae). (Note that in some of the alternative names these taxa have been raised to the level of family as mentioned earlier.)

There are not many crop species within the Mimosoideae and the Caesalpinioideae and the description of the flower which follows is based on that found in the Lotoideae (Fig. 21).

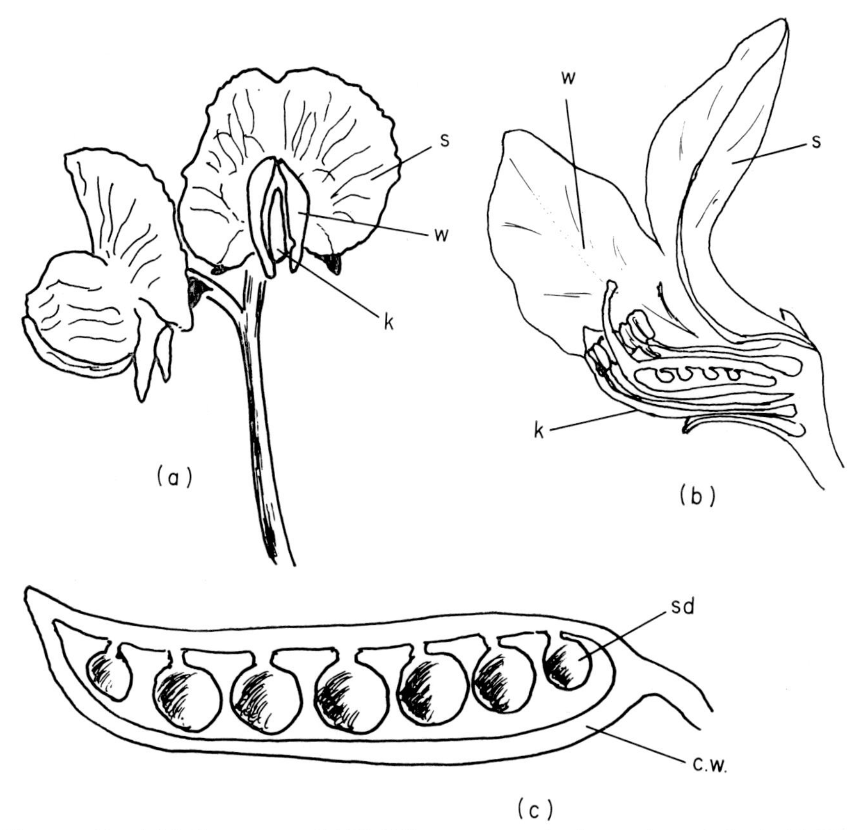

Fig. 21. Pea (a) inflorescence. (b) Half flower: s, standard; w, wings; k, keel. (c) Ripe pod: sd, seed; c.w., carpellary wall.

The flowers are usually large and showy, and produce nectar. These features along with the marked zygomorphy are adaptations to ensure that insect pollination (entomophily) takes place. The flowers themselves have the parts in fives. The sepals are arranged such that the oldest is anterior, which is unusual. This calyx with the odd sepal anterior is zygomorphic with two arrangements found in the subfamily. In the first the two posterior sepals are joined loosely along their contiguous margins, while the three anterior sepals are likewise loosely joined together. This pattern gives a calyx of two parts. The second arrangement of sepals that is found results in the calyx having the form of a tube with five distinct marginal teeth. However, the teeth are not symmetrically disposed around the rim of the tube, so the zygomorphy is apparent. Alternating with the five sepals there are five petals. Each of the petals is distinctive. There is a very large posterior petal called the *standard* or *vexillum*, with two lateral petals often tapering to a narrow base but having a prominent terminal region. These lateral petals are termed *wings* or *alae*. The two anterior petals are loosely joined and together they make a boat-shaped structure within which lie the essential parts of the flower. The anterior petals make up the *keel* or *carina*.

The stamens are ten in number in two whorls each of five. The filaments are connate even though in many cases five of the stamens are long and the other five short. In some species the stamens remain free. Connation of the filaments can mean either that all ten stamens are involved to give a distinct filament tube, or that the upper stamen is free, so that the stamen arrangement is stated to be 9+1. When all ten filaments are joined the flower is said to be *monadelphous*, and where the stamens are arranged 9+1 the flower is *diadelphous*.

A single, superior carpel surmounts the receptacle and its ventral side is prominently uppermost. The ovules which are anatropous-alternate, may be from one to many (up to 15). At maturity the fruit is most often a legume, but single-seeded fruits might not dehisce or instead of dehiscing along the sutures a line of dehiscence develops transversely just below the apex to allow the cap region of the fruit to fall away. Transverse dehiscence lines are formed in some multiseeded fruits to give a lomentum. Drupes occur rarely. The legume might be inflated or twisted into a compact spiral.

In wild species the seeds are hard and resistant to abrasion. If eaten the seed can withstand passage through the gut and this serves as a long-distance dispersal mechanism. The extreme hardness of the seed appears to contribute to its longevity, there being authenticated instances of seeds remaining viable for nearly 200 years. Hard seeds do not take up water until the testa has been modified either by abrasion in moving soil or microbial action. Once the testa is abraded water is imbibed and the germination process starts. Some of the cultivated species produce hard seeds and this can be an important factor in establishing the crop.

It was stated earlier that the flower structure is well adapated for entomophily. The association between members of the Leguminosae and insects is not that of a simple attraction of the insect to a showy flower with a supply of nectar.

The stamens and carpel lie within the keel, and in order to effect cross-pollination the pollen must be delivered from the stamens to the abdomen of the visiting insect. Four mechanisms for transferring pollen from within the keel to the insect have been recognized:

1. When the insect lands on the flower and enters the corolla tube it brushes past sensitive regions of the wings and the flower releases pollen explosively. On release of the pollen the flower is said to be 'tripped' and there is no further possibility of pollen being shed on to an insect at any subsequent visit. This type of mechanism is found in lucerne.

2. The stamens and carpel if contained completely within the keel can be extruded by an insect landing on the keel such that the pollen issues from the top of the tightly enclosed keel to dust the abdomen of the insect. The style and stigma follow. This piston-type mechanism is found in lupin, and subsequent insect visits to the flower result in further pollen extrusion.

3. A somewhat similar mechanism to that described in (2) is found in vetch, but instead of the pollen being extruded by a piston-like action of the stamens and style, pollen issues from the tip of the keel as a result of a brushing motion that results as the style, which has brush-like hairs upon it, is swept up the keel cavity. Repeated insect visits are possible and indeed are necessary for the mechanism to work.

4. This fourth type of pollination mechanism is like (2) and (3) inasmuch as the keel encloses the stamens and the style. In this case the stamens and the stigmatic surface issue from the keel when the insect visits the flower, but on its leaving, the stamens and the style return to the keel cavity. Clover shows this type, and repeated insect visits are possible.

Even with such well-developed morphological adaptation to ensure cross-pollination the Leguminosae possess a number of species in which there is exhibited genetic self-incompatibility. In those species which have been examined this incompatibility is of the gametic type, and its occurrence can be of some considerable importance in the production of seed in red clover crops since there is an obligate cross-fertilization involved. In contrast to these cross-pollinating and cross-fertilization mechanisms a few species are more or less, or completely, self-fertile, some even shedding pollen within the unopened bud so giving rise to what is effectively cleistogamy.

The Fixation of Atmospheric Nitrogen by Leguminous Plants

We have mentioned on two occasions that the nitrogen content of leguminous plants was, on average, twice that of other seed plants. It has been known since

Roman times that clovers could act as soil ameliorators, but the reason for this did not become apparent until the 19th century.

When a leguminous plant is removed from the soil it is found that the roots usually have present on them small swellings with much contorted surfaces. It is often thought that the possession of these swellings, or *nodules*, is a universal occurrence in the family but a survey of 1200 species disclosed that in 133 the roots were non-nodulated. It is possible that the investigators, at the time at which they carried out their sampling, missed noting the presence of these nodules on a particular species. Tropical tree species of legumes have nodules of limited life span and unless an observation were made at a time when the plant was carrying them it would be scored as not possessing nodules. However, other surveys have been conducted and not only has it been shown that the presence of nodules is not universal but it has also been demonstrated that the Lotoideae are regularly nodulated while the Mimosoideae have only about one-third of the species nodulated. The Caesalpinioideae have a degree of nodulation intermediate between that of the other subfamilies.

The distribution of nodules on the root system is not random. Two main types of distribution are found. In the first the nodules are large, few in number and concentrated towards the crown of the root system. In the other, small nodules are dispersed more or less evenly over the whole of the root system with the smaller laterals as likely to be carrying nodules as the major roots. Under certain soil conditions, where there is a deficiency of nitrogen, plants of the first type are found to be robust, and a healthy green colour. Plants with nodules distributed according to the second pattern are most likely not to be very robust, and may even be slightly chlorotic.

A much more telling observation can be made when leguminous plants are grown in a medium which does not contain any combined nitrogen. It is possible to grow plants without nodules, and when this is done these non-nodulated plants cannot flourish in a nitrogen-free medium, whereas the nodulated plants can grow, some almost normally but others poorly.

If a large nodule from a plant which is growing well in a nitrogen-free medium is removed and cut through the middle it is found that the centre of the nodule is bright red. If the same is done with a small nodule the centre of it is seen not to be red but pale cream or creamy-green. The possession by plants of nodules which have red interiors is associated with a capacity to utilize atmospheric nitrogen whereas pale nodules seem not to be able to fix sufficient nitrogen to support the plant.

Comparing the growth and nitrogen content of nodulated plants with that of non-nodulated plants, both growing in a medium free of combined nitrogen, does not prove directly that the nodulated plant was fixing atmospheric nitrogen, but the inference is strong. Direct proof that nitrogen was fixed by the nodulated plant did not come until the heavy isotope of nitrogen, ^{15}N, became available. This isotope can be observed and measured by mass spectrometry, and by exposing nodulated plants to an atmosphere containing a proportion of ^{15}N

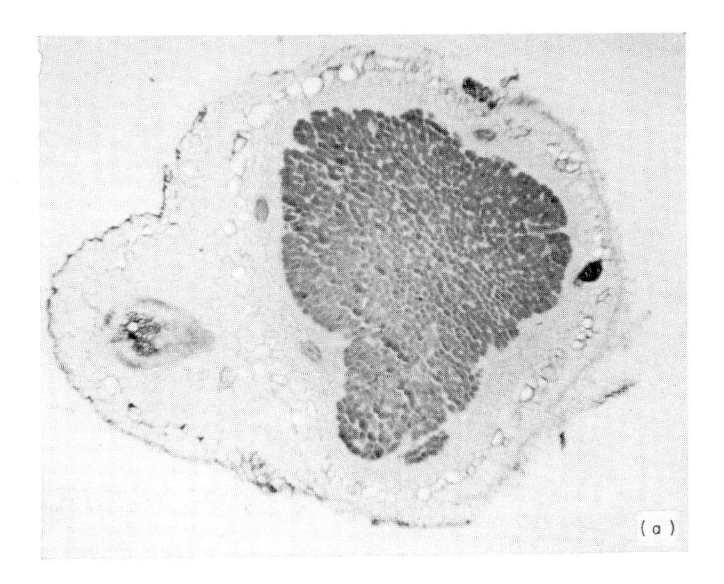

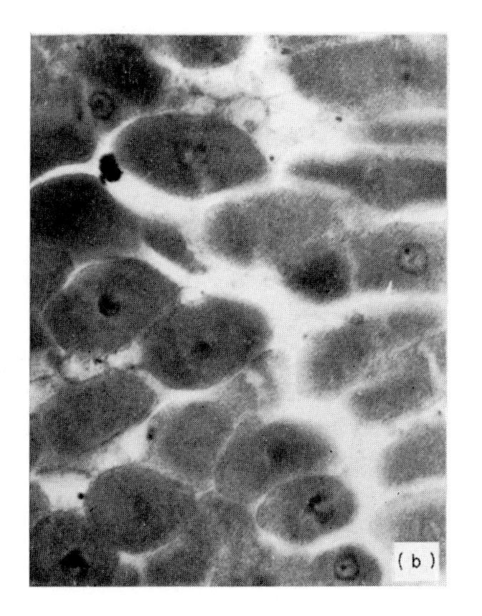

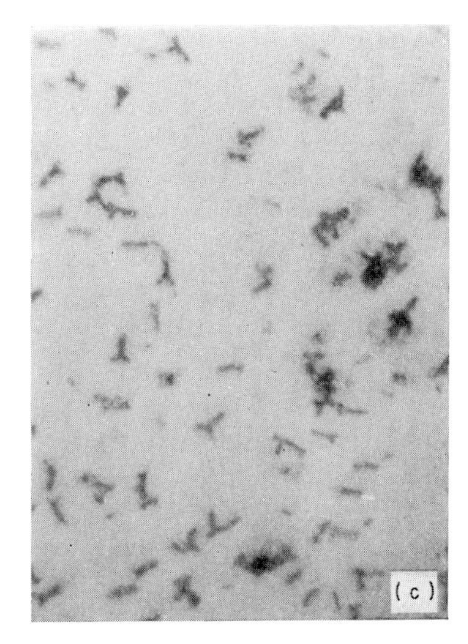

Fig. 22. (a) T.S. of root and nodule of Soya bean. (b) High-power view of bacteroid-containing cells of (a). (c) Smear preparation of *Rh. leguminosarum* from broad bean.

and finding the degree of enrichment of the nitrogenous compounds in the plant unequivocal proof of nitrogen fixation was obtained.

Microscopic examination of the nodule shows that it is a modified lateral root with central cells which are either large and filled with bacteria, or small and containing many starch grains (Fig. 22). The bacterium within these central cells is unusual in that it is not seen to be any one of the expected bacterial forms. The cells are rod-shaped or branched and rather uneven, so that they appear as rather lumpy X's and Y's. They are said to be coralloid. It is for this reason that they have been called bacteroids. The bacterium can be isolated from the nodule, and taken into culture by standard bacteriological techniques. Selective media are employed and the organism is like a normal soil bacterium as far as its physiological requirements are concerned, but its growth rate is a little bit less than that of the more vigorous soil bacteria.

In culture the bacteroid form is not evident. The cells are small flagellate rods and when stained by the Gram technique they are red, hence are *Gram negative.* Soil samples are found to contain similar bacteria with respect to morphology, staining and culture characteristics, and it is clear that the bacterium of the nodule is also a bacterium of the soil where it is free living. In the free-living condition it does not fix nitrogen, nor does the plant when it is growing alone. Only when the two organisms are found together in *symbiotic association* is the phenomenon of nitrogen fixation found. The organism has been given the generic name *Rhizobium.*

It was a short step from the isolation of the organism to the deliberate inoculation of roots of leguminous plants grown in an environment where the *Rhizobium* did not occur. *Rhizobia* isolated from one species could always produce nodules on the same species but did not always produce nodules on another species. After a very substantial amount of work it was found that the genus *Rhizobium* could be divided into groups, and the members of any one group could incite the production of nodules on a number of species of legume but on no others.

Sixteen such *cross-inoculation groups* are considered, but when other characters, such as ability to ferment certain sugars and the nature of colony growth are taken into account, only six species have been erected:

Bacterial species	*Representative Leguminous Host*
Rhizobium japonicum (Kirch.) Buch.	Soya bean
Rh. leguminosarum Frank emed. Bal. and Fred	Pea, broad bean
Rh. lupini (Schr.) Ecktr. *et al.*	Lupin
Rh. meliloti Dang.	Sweet clover, lucerne
Rh. phaseoli Dang.	Bean
Rh. trifolii Dang.	Clover

We shall see how these species accord with the taxonomy of the Lotoideae.

The long association between the legume and its symbiont has meant that the host has had a long period to evolve a metabolism which had available an abundant supply of nitrogen. Nowhere in the plant kingdom is there such a wealth of unusual and unique nitrogenous compounds as in the Leguminosae. Other organisms have not encountered this situation and it is not surprising that many of these nitrogenous compounds are poisonous. The Leguminosae is a dangerous family. Amongst the substances met are α-amino acids which are not constituents of proteins, non-α amino acids, seleno-amino acids (amino acids containing selenium), alkaloids, cyanogenetic compounds and nitro-compounds. As well as these simple molecules the Leguminosae are rich in proteins, and these can simply be toxic such as abrin from the precatory pea, one seed of which can kill a man, allergenic, e.g. fabin from broad bean, or haemagglutinins like canavallin from sword bean.

The presence of a toxic compound does not depend simply on the species. Climatic conditions and genetic constitution can determine whether a particular plant will be poisonous. This can be illustrated with reference to the cyanogenetic glycosides found, for example, in clover. The most common of these compounds are linamarin (I), lotaustralin (II), acaciapetalin (III), and vicianin (IV). Under the action of the enzyme linamarase these compounds liberate hydrocyanic acid, HCN, which is toxic to mammals. For any plant to be toxic to the animal which ingests it there must be (1) an adequate level of glycoside, and (2) linamarase. One without the other is not likely to give rise to toxic symptoms.

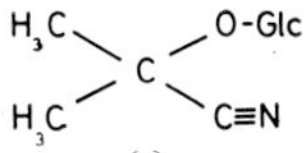

$(H_3C)_2C(O\text{-}Glc)C{\equiv}N$ (I)

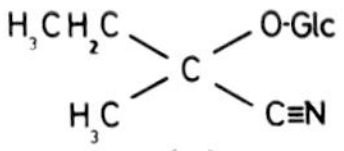

$(H_3CH_2C)(H_3C)C(O\text{-}Glc)C{\equiv}N$ (II)

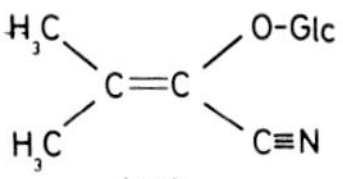

$(H_3C)_2C{=}C(O\text{-}Glc)C{\equiv}N$ (III)

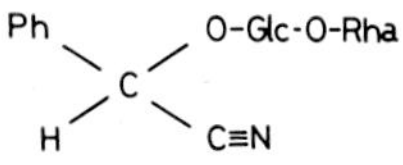

$(Ph)(H)C(O\text{-}Glc\text{-}O\text{-}Rha)C{\equiv}N$ (IV)

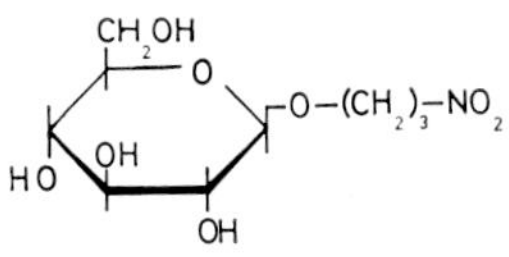

Glucose ring (CH_2OH, HO, OH, OH)$\text{-}O\text{-}(CH_2)_3\text{-}NO_2$ (V)

The level of glycoside that can be present is determined genetically in the first instance, but the stage of growth of the plant and the climatic conditions can influence the genetic potential to a considerable extent.

A rather unusual type of nitrogen-containing compound is found in some of the loco-weeds. It is most unusual for nitrogen to be found in plant products as the nitro group but this is the case in miserotoxin (V). Whether this is toxic *per se* is difficult to assess because in ruminants nitropropanol is liberated in the rumen and this could be the cause of the death of the animal.

It was mentioned earlier that the Leguminosae produce a rather large number of proteins which can affect mammals deleteriously. The first class of such proteins are those which are allergens. Most often the effect that such allergens have on man is to cause discomfort, but in extreme cases of hypersensitivity to the protein the effect can be fatal. Hypersensitivity can arise from frequent recurrent exposure to the allergen and in the case of Mediterranean peoples this can happen with some of the vetches, which produce the protein vicin, and the broad bean which has fabin as an allergen.

Amongst the sword beans, and the lablab bean, as well as some other species, the seeds contain a protein which is able to bring about agglutination of specific blood proteins. Normally such agglutination can only occur when the blood protein is mixed with a specific antiserum, containing the appropriate antibody. These phytohaemagglutinins are almost as specific as the appropriate antibody for distinguishing the blood groups A, B, and O when used at the correct dilution. It is claimed that the plant 'antibodies' can even distinguish subgroups within group A blood types.

In all probability these phytohaemagglutinins have evolved to protect the plant from animal pests, and the ones which are specific for given blood groups represent the fine tuning of evolution. In some Leguminosae there are agglutinins which could be thought of as the whole orchestra, for the effect they have on an animal is drastic and final. A protein of this type is abrin found in the seed of precatory pea (*Abrus precatorius* L.) also known as rosary pea, and crab's eyes. This small, brilliantly coloured black and red pea is strung into beads for religious purposes in many countries. One seed if chewed can be fatal to man. The seed can be swallowed whole without injury, but if chewed the protein abrin is liberated. The protein must be taken directly into the blood stream, and the only way this can happen is if there are open wounds in the mouth or alimentary canal.

Alkaloids are present in many seeds and these can be fatal if ingested. The victims are likely to be children since the seed of the toxic species are superficially similar to those which are daily items of diet, and it is not unusual for children to play at shopkeepers and cooks. In the process of play the seeds are eaten, often with fatal results.

Exploitation of the Leguminosae

A very large number of the Leguminosae are grown as crop plants and the different classes of crop are as follows:

1. Plants grown for their seeds or fruits which are eaten directly by man or his domestic animals: the pulses.

2. Plants grown for their stems and leaves which are eaten green by domestic animals or conserved to be used later: the forage legumes.
3. Seeds harvested to be used primarily in the production of vegetable oil: oil seed crops. (Often the residue left after extraction of the oil can be eaten by man or his animals.)
4. Plants exploited for timber.
5. Plants from which dyes can be extracted.
6. Plants which exude gum (particularly when wounded) which is then collected.
7. Plants grown for a particular chemical compound or groups of compounds which can be used as flavourings, medicines or insecticides.

The exploitation of these groups of legumes and the classification of the family march hand in hand. Most of the pulses belong to a single tribe and most of the forage legumes are members of two tribes, but the species which are grown for their particular chemical component are found throughout the family.

The classification employed below the subfamilies level is not clear-cut, and it is possible, on the basis of the accepted tribal classifications in the Lotoideae, to assign a plant to more than one of the main tribal divisions, because of the equivocal nature of the characters used.

There is a wealth of relatively easily identifiable chemical compounds within the family and there is the possibility of using them taxonomically. Even the growth hormones, the gibberellins, could be used in this way for it has been found that a few are of unique occurrence in some members of this family. While *chemotaxonomy* holds promise in the Leguminosae it is not something which can be undertaken without the expenditure of a considerable amount of time and effort, and to date it has not proven any better than the traditional methods which use morphological features.

Because of the relatively greater importance of the Lotoideae our treatment will emphasise this subfamily. The tribal divisions are erected on the basis of the characteristics of the stamens, the form of the fruit, the nature of the leaves, and features of the ovary. The major tribes (after Taubert) are:

1. Stamens free	
Pinnate leaves	Sophoreae
Simple or palmate leaves	Podalyrieae
2. Stamens monadelphous or diadelphous	
Fruit a lomentum	Hedysareae
Fruit a legume or indehiscent, leaves simple or palmate	
Leaf with three entire leaflets	
Plants herbaceous	Galegeae
Plants shrubby	Genisteae

Leaf with three toothed leaflets	Trifolieae
Leaf with five entire leaflets	Loteae
Leaves pinnate	
Leafstalk ending in bristle	Vicieae
Leafstalk, otherwise, pod dehiscent	Phaseoleae
Leafstalk otherwise, pod indehiscent	Dahlbergieae

In trifoliate species it is not always a simple matter to decide if the leaf is palmate or pinnate, but it is generally considered that if the terminal leaflet is distant from the laterals then the leaf is pinnately compound. If the leaflets are toothed, regardless of whether the terminal leaflet is distant or not, it is advisable to consider such species in the above classification to have palmately compound leaves, and these are therefore members of the Trifolieae.

The pulses

Two tribes have provided man with the greater part of his pulse crops—the Phaseoleae and the Vicieae. In all cases the seeds are relatively large, being more than 4 mm in diameter, and in some instances the seed is very large. Those members of the tribes which are most successfully grown as pulses have seeds rich in carbohydrates and with the high level of protein associated with legume seeds they provide man with a more balanced diet than wheat. In many parts of the world the legumes provide the local population with its main source of protein, and while this is not considered to be of equivalent quality to animal protein, being deficient in some of the essential amino acids, the legume protein does represent a better class of protein than is normally encountered amongst plants.

A few of the genera produce seeds with a high oil content, and while these can be eaten, the oil is usually extracted leaving behind an enriched meal. This is the case with soya bean, and the meal remaining after the removal of the oil is so much enriched that defatted soya meal (or flour) has a protein content equalling that of animal products.

The tribe Phaseoleae might be called the bean tribe, but in using the word bean it is essential to be specific and for certainty the Latin binomial should be employed. Bean has a similar local usage to the word corn, and means that pulse crop grown locally for consumption. Amongst the genera of the Phaseoleae that have given us beans are *Phaseolus*, *Glycine*, *Canavalia*, *Cajanus*, *Vigna*, and *Lablab*. In terms of world trade the soya bean is undoubtedly the most important of the world's pulses, and its importance increases annually. It is a most valuable export item in the economy of the United States of America and in 1974 the value of exports was worth $3.10.[9]

For a pulse crop to be successful a good fruit set is necessary, and as well as being determined by climatic conditions, this setting of fruit is conditional on pollination and the type of breeding system encountered in a species. Those species which depend on cross pollination are not as reliable as those in which self-pollination associated with self-compatibility is the rule. In some species pollen is shed within the unopened flower and these are particularly valuable. Where cross-pollination depends on a specific insect pollinator, then if it is absent at the time the flowers reach anthesis, crop potential is reduced. Furthermore the uniformity that results from generations of inbreeding allows for an easier harvesting of the crop.

The genus *Phaseolus*

Most of the members of this genus are annual herbs, though in a few the root stock can become woody and serve as an overwintering organ. *Phaseolus* is distinguished from the other members of the tribe by having the keel coiled in the opened flower. This feature can be most marked.

This is a large genus with between 150 and 250 species, distributed throughout the tropics and subtropics, with recognized Old and New World groupings. About 12 species are considered to be of major economic importance and the characters used to recognize these are flower colour (yellow or not yellow), acuteness or otherwise of the leaflets, size and shape of seeds (flat and large or small and rounded), and size of the calyx (large or small).

Seeds of this genus germinate at relatively high temperatures, and this limits the possibility of growing them in temperate latitudes. If plants of this genus are to be grown at higher latitudes than those of its natural distribution, the seeds must be sown late in the season when soil temperatures are relatively high. Germination can be either hypogeal or epigeal, and is normally a species character. The radicle is robust and grows rapidly to anchor the seedling firmly in the ground but radicle growth soon declines and the main root in some species is truncated. There is prolific development of lateral roots and a propensity to form adventitious roots from both the hypocotyl and epicotyl if these are buried in soil.

In those species which exhibit epigeal germination the cotyledons are carried above ground by the extending hypocotyl to as much as 10 cm above the soil surface, and while they turn green they do not expand to contribute much from their photosynthetic activities. The epicotyl is distinct and the first leaves produced arise together at what almost amounts to a common node. These leaves are unifoliate with long petioles and small stipules. Subsequent leaves are trifoliate with the terminal leaflet distant from the laterals. As well as possessing stipules, these leaves have at the bases of each of the leaflet stalks, homologous structures termed stipels. In all instances there are well-developed pulvini at the bases of the petioles and the leaflet stalks. At excessively high temperatures the laminas are moved so that they are disposed vertically upwards, the blades thus presenting their edges to the sun. This is a remarkably efficient adaptation to minimize the heat load on the leaf. At low temperatures and at night the leaflets adopt a vertically downwards aspect, so reducing radiant heat losses. The axis can develop monopodially or sympodially. In the former case the

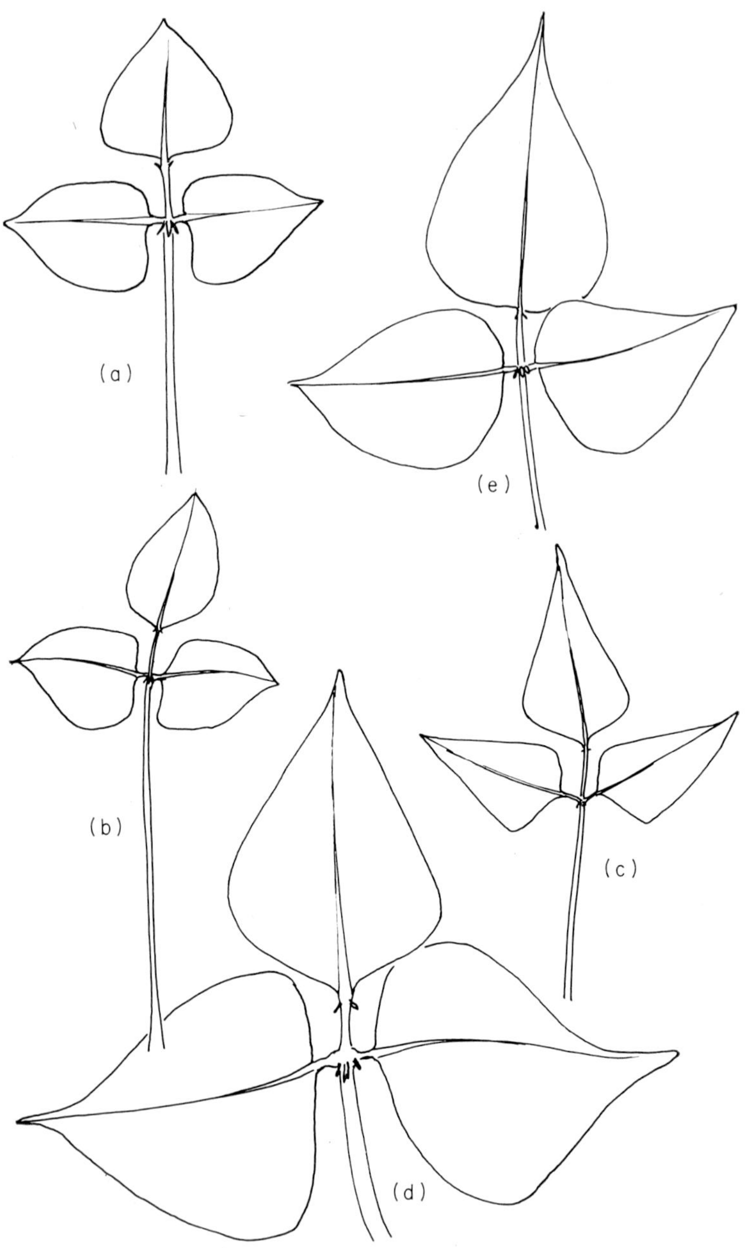

Fig. 23. Leaf profiles of (a) *Ph. aureus* (b) *Ph. mungo* (c) *Ph. lunatus* (d) *Ph. coccineus* (e) *Ph. vulgaris*.

internodes are long and the plant is a straggling or creeping herb. Those plants which have sympodial growth produce from four to eight internodes on the main axis and then the axillary buds begin to grow out to give a bushy herb quite different from the climbers. The plants are softly pubescent and the stems and petioles are variously grooved.

The flowers are produced in racemes with up to 12 flowers. In the climbers these racemes are axillary, and within a genus cultivars can be grouped according to where these inflorescences arise on the axis. Early cultivars produce inflorescences at the lower nodes (beginning at about node 11) while the late cultivars do not produce flowers until perhaps the thirtieth node. Bush beans produce terminal inflorescences and as they are produced branching is promoted. It is difficult to make out a pattern in this branching but it is unusual for the branches to produce more than five nodes before terminating in an inflorescence. Both climbing and bush phenotypes occur within a species, and while the phenotype is fundamentally determined by the genotype environmental modification does occur, especially due to long daylength.

Flowering in *Phaseolus* is either under day-length control, the plants being short-day plants, or the plant is day-neutral.

The commercially important species of *Phaseolus* have originated in two centres—the Old World in eastern Asia, and the New World in the Central Isthmus. All the Old World species have yellow flowers, whereas the others have white, lilac, pale purple, or scarlet-red flowers. The Old World species are described first, below:

OLD WORLD

Phaseolus aconitifolius Jacq., mat bean. Germination: epigeal. Habit: a much-branched trailing or climbing annual, noticeably hairy. Leaves: alternate, trifoliate carried on grooved petioles 5–10 cm long; leaflets up to 8 cm long, divided, so giving the species its name. Inflorescences: axillary carrying several small yellow flowers; peduncle equalling length of petiole; self-compatible, self-pollinated. Pods: small, brown containing up to 9 seeds, these being 5 mm long, yellow, black or parti-coloured, with a linear white hilum.

Ph. angularis (Willd.) Wight, adzuki bean. Germination: hypogeal. Habit: erect bushy plant up to 75 cm tall but straggling cultivars occur, though this may be a response to daylength. Leaves: alternate, trifioliate with long petioles; leaflets entire, ovate up to 9 cm. Inflorescences: axillary, carrying between six and 12 brilliantly yellow flowers; peduncle shorter than petiole; self-compatible, self-pollinated. Pods: up to 12 cm long with up to 12 seeds present. Seeds longer than broad, not flattened, self-coloured red-brown to black, but lighter forms exist, the hilum being prominent and white.

Ph. aureus Roxb., mung bean, green or golden gram. Germination: epigeal. Habit: initially this annual grows as an erect bush form but the upper part of the plant becomes twining, the total height of the stem reaching to 1.5 m. The plant is hairy. Leaves: alternate, trifoliate with long petioles; the leaflets are ovate to 12 cm long. Inflorescences: axillary with the peduncle up to 12 cm but shorter than the petioles; many-flowered, the flowers being large, yellow; self-compatible, self-pollinated. Pods: grey to brown containing up to 15 seeds which are deep olive-green, small and rounded; the hilum is white and discoid.

Ph. mungo L., black gram, urd. Germination: epigeal. Habit: branched hairy annual, hairs reddish-brown; occasionally exhibiting twining character but never as robust or prominently twining as the previous. Leaves: alternate, trifoliate with long petioles, leaflets narrower than in *Ph. aureus*, approaching lanceolate. Inflorescences: axillary, with short hairy peduncle which may branch; up to five flowers at the end of the peduncle; flowers moderate size, pale yellow; pollination occurs in the bud and inbreeding is obligatory. Pods: buff to dark brown covered in hairs, with up to 10 small seeds, usually black; hilum white, concave. Recently, some authors have considered that these, and the other, Old World species should be assigned not to *Phaseolus* but to other genera, in particular, *Vigna*. Splitting *Phaseolus* in this way may be justified on the basis of the lack of gene-exchange between the two geographical groupings.

NEW WORLD

Ph. coccineus L., scarlet runner bean. Germination: hypogeal. Habit: a climbing perennial, but normally treated as an annual. The gene for bush habit exists and there are cultivars which are homozygous for this character. The long pods of this species trail on the ground if the plant is too dwarf. The perennial habit depends on the production of a tuberous rootstock which can be eaten, and in temperate regions this rootstock may be uplifted in autumn and overwintered indoors, to be planted in the following spring. Leaves: alternate trifoliate with leaflets to 20 cm. Inflorescences: axillary with peduncles longer than the leaves carrying many flowers, only some of which produce fruit. The flowers are large and showy, usually scarlet red, but sometimes pale. Pods: very large and flattened, sutures prominent and wrinkled, can reach to 50 cm; seeds large to 2.5 cm, flattened, and slightly curved, particoloured purple-red and dark brown; hilum black; cross-pollinated.

Ph. lunatus L., Lima bean. Germination: epigeal. Habit: variable with tall pole forms approaching that of the previous species, and dwarf bush type like *Ph. vulgaris* (q.v.). The plants are hairy. Leaves: alternate, trifoliate with petioles up to 20 cm. The leaflets are rhomboidal with acuminate tips. Inflorescences: axillary, with a peduncle as long as the petiole, carrying many flowers in clusters; flowers small, greenish-cream; self-compatible, self-pollination but considerable outbreeding can take place. Pods: small with 2–4 seeds. Seeds of two types occur within the species—flat bolster forms up to 3 cm long and smaller rounded tupes, at most 1 cm long. The flattened seeds types are those often called butter beans or Madagascar beans, and also the bean eaten in the United States in the dish called succotash. The smaller beans are the navy beans of commerce which are most often prepared in a tomato sauce and canned.

Some seeds contain cyanogenetic glycosides, which are destroyed in cooking. Seeds are usually creamy-white, but cultivars with coloured seeds exist.

Some confusion surrounds the taxonomy of this species. The large seeded types have been given specific rank as *Ph. limensis* Macf. while the species *Ph. lunatus* was restricted to the small seeded types which were called colloquially the Sieva bean. It is now accepted that the two distinct types of seed are found within the same species.

Ph. vulgaris L., kidney or French bean. Germination: epigeal. Habit: variable, cultivars showing the bush form and the pole form but often bush types will exhibit elongated internodes and the twining habit in the upper part of the axis. This latter type of growth is seen when the species is grown at high latitudes and under long days; plant somewhat hairy. Leaves: alternate, trifoliate, hairy, petiole long and furrowed on upper surface; leaflets ovate, up to 25 cm long. Inflorescence: terminal or axillary. In the bush forms, inflorescences are frequently terminal, but in the case of those cultivars which have an indeterminate growth pattern the inflorescences are axillary; flowers few towards the ends of peduncles, shorter than the leaves, moderately large, coloured cream, white or lilac; pollen shed within the flower as the bud opens, leading to inbreeding. Pods: narrow, long, up to 20 cm with as many as eight seeds per pod; seeds rounded and plump, slightly curved, hilum white, varying in colour from cream to russet-brown with a dark hilum.

The groups of cultivars which are recognized are erected on the type of pod and seed produced.

In the string beans the vascular strands in the sutures are well developed, and if the pod is allowed to develop filled seed before harvesting, these strands must be removed before cooking. There is a mutant form in which the vascular strands are not prominent and the pod is more inflated. The carpellary wall is more succulent, and in these the pod can be snapped transversely. Types with this pod are called snap or wax beans.

Those varieties with small white or cream seeds make up the true haricot beans, and they are much prized for their flavour.

In this treatment of the members of the genus *Phaseolus* the individual flowers have not been described in detail. This has been because there is not much fundamental difference in the forms of the flowers, though it should be remarked that in these beans the flowers are subtended by two bracts that appear like a second supernumerary calyx. These bracts can be diagnostic. The extent of spiralling of the keel and the style within it are also helpful in identifying a species.

Important as the *Phaseoli* are, their value as crops cannot compare with the next species that we shall consider—the soya bean.

Glycine max (L.) Merr., soya bean: this genus has a more northerly distribution than the previous, and the cultivated soya bean is not found in the wild, unlike some of the *Phaseoli* which may, however, be escapees from cultivation, The crop plant is thought to have been derived from *G. ussuriensis* Regel and Maack which occurs in eastern Asia. We think of the crop as belonging to the area of north-east China, Korea and Japan. It is only within the last few decades that it has been planted extensively elsewhere. It is being developed as a possible crop species to be grown in the United Kingdom, though it is recognized as being subtropical, primarily because of its response to daylength and temperature.

During the early period of investigation, after the introduction of many strains into the United States, it was considered that the species contained a number of different maturity groups. These were designated by Roman numerals, up to X, the larger the numeral the later the maturity group. Since then the investigations have disclosed very early maturing groups designated O and OO, and there may even possibly be a group OOO. While the species is considered to be a type requiring a short day, there are so-called long-day strains. Possibly the maturity groups represent forms which are all short-day types, but each having a different critical daylength. The OO types will flower under conditions where the daylength is long, exceeding 16 h, but their behaviour is that of a true short-day plant. The maturity group X will only flower if the day-length is short, not exceeding ten hours. Temperature during the growing period contributes to the capacity to flower, and with the early maturing types low temperatures are tolerated. It is within the early maturity groups (OO–II) that cultivars are being sought which might be grown towards the more northerly part of the plant's range.

Another important consideration when introducing this plant into new areas is that of providing the appropriate strain of bacterium for the production of effective nodules. The species of *Rhizobium* that can enter into an effective symbiosis with *Glycine* is *Rh. japonicum*. This is absent from the soils of Europe and America and if the crop has to be grown without the need for large doses of nitrogenous fertilizer, it is necessary to inoculate the seed before it is sown. This is done by purchasing from a recognized source cultures of tested strains of *Rhizobium* and preparing a suspension in skimmed milk. The seeds are coated with this suspension and sown immediately. If sowing cannot take place soon after inoculation the seeds need to be stored in a cool, dark, humid place to prevent death of the bacterium by desiccation and exposure to sunlight.

The seed is rich in oil, some cultivars containing more than 25%, and this, along with protein levels as high as 50%, makes this seed one of the most valuable food materials available.

The oil that is obtained is a high-quality vegetable oil which can be used in the manufacture of both food stuffs and industrial products. It is the most important oil in the production of margarine in many parts of the world. Its constitution with respect to unsaturated fatty acids makes it a drying oil, so it can be used in paints and varnishes. The oil as expressed contains a number of fat-soluble substances, amongst which are vitamin E and plant sterols. The level of sterol in the oil is sufficient to warrant its extraction to be used by the pharmaceutical industry in the preparation of steroid hormones, and soya bean oil is the biggest single source of plant sterols used in this way.

The residue after extraction of the oil is rich in protein and is a most valuable animal food. Both soya meal (flour) and defatted soya flour can be used to supplement wheat flour to give to a loaf a greater nutritional value than the standard wheat loaf and these protein-rich flours can also be processed to a material which is texturized, and can be used as a meat extender. Meat can be 'extended' 30% without a noticeable change in flavour, and in the case of minced beef without any change in texture or consistency.

Soya bean can be grown as a forage legume, and it is also grown as a green manure. This last practice is employed with many legumes and consists of planting a legume, allowing it to grow vigorously and ploughing it into the soil while still green. This leads to soil improvement and is good husbandry where animal manure is not available for adding organic matter to the soil. The fixed nitrogen can be released to the subsequent crop, also improving yields.

Germination in soya is epigeal, and like all members of the Phaseoleae with this type of germination the cotyledons do not act as significant photosynthesizing units. Unifoliate leaves are produced, to be followed by the adult trifoliate leaf, though leaves with five pinnae are not uncommon. The leaves are markedly hairy and have small lanceolate stipules with insignificant stipels at the bases of the leaflets. The petiole is long, without grooves and moderately fine.

The main axis grows upright to over a metre when vigorous, and in the most popular cultivars secondary and tertiary branching is pronounced, with the branches being erect ascendent. Some of the older cultivars, now less popular, exhibited a prostrate habit. The stems are markedly hairy.

The flowers, which are small, are carried in axillary racemes arising from the base of the plant upwards. There can be as many as 15 flowers in each inflorescence. The species is self-compatible and self-pollination is the rule, but in the field a small amount of outbreeding is found, there being reports of up to 2% cross-pollination, although there has not been an intensive study of the breeding system under agricultural conditions.

After fertilization pod development is rapid; the pod at maturity varies in colour from pale grey to black and contains in most cases three seeds. The seeds themselves can be cream, green, brown or black and are about 1 cm long slightly curved and round.

Throughout the tropics other species of the Phaseoleae are found to be important locally. Few, if any, of these species ever enter world commerce. They are nearly all pole forms, and may be either annual or perennial. The description of the plant habit that has been given for *Phaseolus* and *Glycine* could almost equally apply to any of the following, the differences which exist being quantitative not qualitative.

Cajanus cajan (L.) Millsp., pigeon pea.
Canavalia ensiformis (L.) D.C., jack bean.
C. gladiata (Jacq.) D.C., sword bean.
Dolichos uniflorus Lam., horsegram.

Lablab niger Medik., lablab bean.
Pachyrhizus erosus (L.) Urban, yam bean.
P. tuberosus (Lam.) Spreng., yam bean.
Psophocarpus tetragonolobus (L.) C.
Vigna unguiculata (L.) Walp., cow pea. This can be thought of as an aggregate species since *V. sinensis* (L.) Savi ex Hassk. and *V. sesquipedalis* (L.) Fruw. are splits from the original wild species, assigned to cultivated forms. If the cultivated forms are thought to be separate from the wild progenitor then *V. sinensis* is the preferred name.
Voandzeia subterranea (L.) Thou., Bambara groundnut.

This last species is distinctly different from the standard habit we have assigned to the Phaseoleae and it is worth mentioning its distinctive features. In its habit it is a creeping plant rooting at the nodes, and in many ways is similar to white clover (q.v.). The two- to three-flowered inflorescences are axillary and initially have short peduncles. After fertilization the peduncle elongates and grows positively geotropically to bury the fruit in the soil where maturation of the seed takes place.

The Vicieae

What the Phaseoleae are to the tropics so are the Vicieae to the temperate regions. Some species of this tribe are warm climate plants but the single most important pulse crop of the United Kingdom, peas, belong to this tribe. It may be that with continued improvement the pea might be able to produce amounts of protein per acre comparable to soya bean.

Pisum sativum L., pea: the taxonomy of this species can lead to some difficulty. Within the genus nine species are recognized and the original Linnaean species referred to the garden pea with white flowers and green or yellow seeds. Linnaeus considered that the field pea with its purple flower and brown mottled seeds was a distinct species, and he called it *P. arvense*, the field pea. There is no justification for considering the two cultivated types as belonging to separate species and in our treatment all peas grown as pulses will be allocated to *P. sativum*.

Peas are pre-eminently a cool-climate crop. Experimental work has shown that if the plants are subjected for prolonged periods to temperatures in excess of 25 °C they will die. On the other hand they will tolerate being frozen, and on being thawed will recover to continue their growth. The plants are sensitive to daylength but do not require a specific daylength to flower and there is a complex interaction between daylength and temperature, and the flowering behaviour. Briefly, the longer the day and the lower the temperature, the more rapid is the production of flowers with respect to the node at which the first inflorescence is formed. Since low temperatures reduce the rate of growth the benefit of having the plants flower at a lower node may be outweighed by their taking longer to reach that particular node in their development.

Germination in pea is hypogeal, and is rapid at temperatures above 15° C. The epicotyl elongates but may or may not bring the first node above soil level. At this node there is a scale leaf which consists of modified stipules with little hint of the petiole, and the lamina is completely absent. The second leaf like the first is a scale leaf but in this one there is greater development of the stipules and the petiole is obvious as a small bristle somewhat flattened at the base. In some varieties the terminal portion of this stipule homologue can be blade-like.

After the third leaf has been produced there is progressive development towards the adult form which has prominent stipules, and a number of pairs of leaflets. The terminal portion of the leaf possesses tendrils which are seen to be the modified distal leaflets. It is considered that the very large stipules are the main photosynthetic organs, but since the leaflets are not inconsequential and the stem is green all parts of the plant can photosynthesize and must be thought of as contributing to the final yield. Very recently mutant forms in which the laminar portion of the leaf is reduced have been tested for productivity and these have been found not to differ from the normal pea with respect to production. Still other mutants, in which as well as loss of the leaflets there is loss of the stipules, have been tested against the normal and in one 'leafless' variety a yield of 50 cwt of peas per acre (approx. 6 tonnes/hectare) was obtained. These 'leafless' forms have the great advantage of being much more easily harvested in the green state, for the users of vining peas, i.e. for quick freezing, rapid dehydration, or canning of fresh peas.

The inflorescences are axillary racemes with up to ten flowers but the commercial cultivars are two- or sometimes three-flowered. Three distinct maturity types are found, depending at which node the first inflorescence is produced. Early varieties form the first inflorescences at the ninth or tenth nodes, mid-season varieties at about the twentieth, and lates after the thirtieth node. Flowers are large and showy, either white or purple. Pods can be large, up to 20 cm long by 6 cm wide, but in varieties grown primarily for their seed the pod is narrower. The pods can contain as many as ten seeds, which may be round or wrinkled, and in the white flowered forms yellow or green. The purple-flowered varieties have mottled seeds brown on purply-black. Not all peas are grown for their seeds; a few—those with a gene for absence of the sclerenchymatous lining of the pericarp—are grown for the pod, which is cooked and eaten like those of French bean; these are the sugar peas. Peas are self-compatible, and since the pollen is shed in the bud the species is inbreeding.

Faba vulgaris Moench., the broad bean, also known as the horse or tic bean, was the only bean grown in Europe prior to the introduction of *Phaseolus* from the New World. In turn it was exported to the Americas after 1492 and has become one of the crop species grown on the Western continents. This plant is also known as *Vicia faba* L. and while it has recognized affinities with the genus *Vicia*, the vetches, it is probably better to separate it as the genus *Faba*, which is monospecific.

Broad beans are very variable and many different subspecies and varieties are recognized. Indeed subvarieties have been named but these last taxa are isolated on the basis of their geographical origins as crop types. The subspecies to which the main cultivated forms belong is the *eufaba* and the divisions of this are given below. *Eufaba* is distinguished from the other subspecies, *paucijuga*, by having more pairs of leaflets (3–4). The subdivisions of *eufaba* are indicated below:

Table 3

Faba vulgaris Moench. *eufaba*: varieties (after Hector)

	Seeds		
Pods	Small black bolster (var. *minor*)	Medium flat bolster (var. *Equina*)	Large flat (var. *major*)
tender	*tenuis* (Asia, Mediterranean)	*rugosa* (Asia Mediterranean)	*clausa* (Mediterranean)
coarse	*rigida* (Asia, Europe)	*reticulata* (Europe, Japan/China)	*dehiscens* (Europe, Japan)

Germination in the broad bean is hypogeal and, as in pea, the first two leaves which are produced on the main stem are modified, being two scale-like structures derived from the stipules. Growth is rapid and the stem is sharply angled and square in section. Sometimes wings are produced along the angles of the stem. The internodes are hollow and at the nodes, again like pea, more than one axillary bud is present, although only one bud develops if branching takes place. Normally the buds in the axils of the cotyledon develop so that the mature plant has three distinct axes. Those which arise from the cotyledonary node are like the main axis with respect to the nature of the leaves produced at the basal nodes. Sometimes up to five more axes can be produced but this is not seen in the cultivars grown in the more intensive agricultural situations. When axillary buds grow out, the basal leaves on them need not necessarily be modified. The adult leaf has up to four pairs of leaflets 5 cm long by 2.5 cm broad but in conditions of high fertility the leaflets can be twice as large. The leaf ends in a small bristle which does not act as a tendril. The stipules are about one-third the size of the leaflets and have dentate margins, and often have a brown spot just below their centre.

In comparison to the erect, non-twining broad bean the vetches approach pea in their habit. Members of the genus *Vicia* are weak climbing plants with straggling stems held upright by tendrils, developed from the terminal leaflets of a pinnatedly compound leaf, attaching to a support. The vetches also differ from broad bean in the way in which the inflorescences are carried, and in the colour of their flowers.

Flowers in broad bean are carried on axillary racemes which have short peduncles and in an inflorescence there are from two to six flowers. The flowers are large and showy, mainly white, though the wings have a black blotch. The calyx is strongly tubular and the long bases of the petals form a conspicuous corolla tube. The flowers are sweetly scented and produce abundant nectar. In all other respects the flower is typically papilionaceous.

Pollination is strictly entomophilous and though the bean is self-compatible much cross-pollination takes place; indeed it seems that under field conditions self-pollination seldom occurs. Fruit set in broad beans is unpredictable and in order to have high levels of fruit set it is necessary to have suitable pollinators present at the time of anthesis. The best pollinator for the broad bean is the bumble bee; its heavy body opens the flowers easily, and with its short proboscis the insect requires to penetrate the flower to reach the nectar. Honey bees are good pollinators, but are more intelligent, and will rob the nectar by cutting through the calyx and corolla tubes. On a still, warm day it is possible to hear the bees' mandibles cutting through the turgid plant tissue.

The pods are various, and as the subvarieties' names indicate, can be sclerenchymatous, coarse, or less so, or tender. Within the pod there are 3–8 seeds from 1–2.5 cm, round or flat, black, brown or white.

The root systems of crop plants are seldom considered in detail and while most of the *Phaseoli* have tap roots which are robust, seldom does the tap root in that genus penetrate deeply into the soil. In broad bean the tap root is vigorous throughout the life of the plant and continues to grow downwards to depths in excess of one metre. The wellbeing of the plant seems to depend on the continued growth of the tap root, for yields are directly related to the length of the tap root. The plant is also very high in minerals. The percentage ash of broad bean is about twice that of other species, and the ash is particularly rich in calcium. These two facts mean that the broad bean is a plant of rich deep soils, and it does not prosper on shallow impoverished podsols. Since the seed is high in protein its value as a crop is high, but it is unpredictable in its pollination and its growth. Cultivars are heterogeneous, since outbreeding is the rule, resulting in a further detrimental feature to detract from its potential usefulness as a crop plant.

The vetches, species of the genus *Vicia*, were important crops in the 19th century and earlier, but they have been supplanted by peas, and broad beans. In habit the most widely grown, *Vicia sativa* L., common vetch or tare, resembles pea though the stem is much branched. Its seeds are small and the yield does not compare with that of the modern varieties of pea. It is easy to understand the reasons for its decline as a crop.

Within the tropics and subtropics there are two important pulses which are members of the Vicieae: *Cicer arietinum* L., the chick pea or gram, and *Lens esculenta* Moench., the lentil. *Cicer arietinum* is an important pulse in places as diverse as Italy and India, and in the latter country is the most important leguminous plant in the diet of the local population.

Germination in this species is hypogeal and the first two leaves which are formed resemble those of pea in that they are no more than small modified stipules. Subsequently the leaves are pinnate with a large number of pairs of small leaflets. The stipules are as large as the leaflets which are from about

1–2 cm long. The whole plant is spreading and much branched and looks very untidy. The flowers are normally solitary in the axils carried on a contorted peduncle, and are characteristically papilionaceous. At maturity the fruit is slightly inflated, containing one or two large seeds. Flowering in this species is extended and plants are found with flowers in bud, ripe fruit and all stages between. This behaviour is undesirable if the plant is mechanically harvested but under the conditions of subsistence agriculture is of considerable value. Yields are not high even with sequential harvesting of the fruit as it ripens (average yields about 0.5 tonne per hectare). Part of the explanation for this low yield might be in the pollination of the flowers. The flowers open only during daylight and normally this opening can take place on two successive days. The anthers are carried above the stigmatic surface and can shed pollen on to the stigma while the flower bud is open. Self-pollination can therefore take place, but only if the days are sunny. Insects visit the flowers and while self-pollination does occur, the compatibility status of the plant is not well known.

Lentils are a crop of drier regions than chick pea, but when grown under favourable water conditions the yields can be moderately high for a sub-tropical plant, reaching over 1 tonne per hectare. In many communities the lentil is the main source of protein, and it is highly regarded for its nutritional value. When split, that is the testa is removed and the cotyledons separated, the product is called dahl which can be made into a very nutritious soup.

Germination is again hypogeal, which we have seen is the norm in the Vicieae. The first-formed leaves are unlike the later-formed adult leaves, again a feature seen in the other members of the vetch tribe.

Lentils are much branched and erect with pinnate leaves, there being up to seven pairs of leaflets each about 1.5 cm long. These leaflets are narrow lanceolate, and correspondingly the stipules are linear. The axillary inflorescences carry 1–4 flowers, each less than 1 cm long and which may be blue, white, or pink. The pod is usually two-seeded, small, not inflated, and broad. The seeds are 5 mm, lens-shaped, with dark testas, but on splitting reveal the characteristic salmon-red cotyledons.

In the Phaseoleae we encountered the most important oil crop in commerce in the soya bean. The last of the tropical pulses which we will consider provides us with another of the world's important oil crops, groundnut. This plant belongs to the tribe Hedysareae, and unlike all the legumes so far considered its fruit is a lomentum which is ripened underground. The oil of groundnut is a high quality non-drying type which can be used for culinary purposes and the residue is an excellent cattle food. In these respects it is similar to soya bean, but it differs from that seed in one important respect. Groundnuts are attractive to eat, whereas soya beans are not the most palatable of foods. This means that the largest producer of groundnuts, India, is not the major exporter of the crop, since most of it is consumed locally.

Arachis hypogea L., groundnut, earthnut, peanut or monkey nut: germination in groundnut is unusual though a somewhat similar type of germination can sometimes be seen in *Phaseolus coccineus*. The hypocotyl extends in this species but the cotyledons are not brought above ground—they are carried to the surface of the soil and when there the hypocotyl ceases its extension. The cotyledons turn green and the epicotyl now begins to elongate.

The leaves have two pairs of leaflets, ovate to obovate, with smooth margins and usually glossy surfaces. The tip of the leaflet is prolonged but it is not abruptly mucronate. The petioles are long, at least half the length of the whole leaf, and may be sparsely hairy, while at the base the somewhat lanceolate stipules are adnate for part of their length to the petiole.

In habit the groundnut shows a separation of vegetative branches and floriferous axes. The vegetative axes are upright and indeterminate and there are produced about eight of these monopodia from the main axis, the cotyledonary buds, and the basal buds of these first-formed shoots, so all these might then be regarded as seminal. At the basal nodes of these monopodia, much reduced branches arise, perhaps of no greater extent than one node, and on these reduced branches inflorescences are produced. The production of floriferous shoots is inimical to the continued production of erect monopodia. Quite considerable variation is shown in the building up of the plant body amongst the various cultivars, but Bunting categorized the variation. To his recognizable groups he gave subspecific rank, but because of the way in which the breeding system in this species operates the distinct morphotypes might be no more than highly inbred lines. However, two good morphotypes can be established. In the first, the main axis produces exclusively vegetative branches. The secondaries produce in sequence two basal branches which are vegetative, then two reduced reproductive axes, then two vegetative shoots, and so on until like the primary axis these secondaries terminate in a 'sterile' region, i.e. the axillary buds do not develop. The second type of plant body consists of a set of primary axes (here we shall consider the shoots from the cotyledonary buds as equivalent to primaries) which from their lowermost nodes develop vegetative secondaries, then a set of reduced flowering secondaries, eventually becoming 'sterile'. The secondaries give rise to a set of basal vegetative tertiaries and upper fertile reduced branches. As already remarked, once flowering is established the development of vegetative axes is reduced.

The flowers are carried on very abbreviated racemes on which there are from one to many flowers per inflorescence. These inflorescences are carried at the basal nodes, and in the case of plants which are earthed up, that is the soil is forced by means of a drill plough around the base of the plant, inflorescences can arise below ground. The individual flowers have a well-developed calyx, which is tubular at its base with the upper teeth prominent. The basal tube can be as long as 5 cm and in fact surrounds a true gynophore. The yellow corolla is distinctive since the petals have exceptionally long basal claws. Within the tube formed by the claws of the petals are contained ten monadelphous stamens of which two, the posterior ones, are sterile, the remaining eight being composed of two groups each of four, and separated by the shapes of their anthers. One group has small globose anthers, while in the other the anthers are oblong. Pollination can take place without the flower opening, and this is certainly the case where the flowers are formed below ground. Above ground the flowers are visisted by insects but it seems as though inbreeding is the main breeding system encountered in this species, the groundnut.

After fertilization the gynophore elongates and carries the developing ovary away from the plant. The gynophore is positively geotropic and soon buries the young fruit, at its tip, in the soil. Below ground the fruit continues its development until ripe. In the early stages of fruit formation the tip of the carpel becomes modified and looks both macroscopically and microscopically like the apex of a root and indeed there are reports of absorptive hairs being produced in the sub-apical region. The seeds begin their development after the fruit has been buried and it is thought that seed formation and final maturation of the whole fruit can only take place in darkness, or at least underground. It is perhaps too much to state that the fruit is negatively phototropic, and this misconception probably arises because Darwin considered that the whole process of burying and ripening of the fruit was under the control of light. It is of interest to note that a fruit which is not buried will turn green and become inflated. It does not ripen seed. When within the soil the ovary usually lies horizontally and when the fruit is ripe there are present two seeds. The seeds are ovoid often with one end sharply truncated, and about 1 cm or more long. The testa, which may be brown, pink, white or red, is thin and papery, and within it is the embryo with two massive cotyledons. The plumule and radicle are well developed.

As will be apparent from our description of the pulse crops the species of importance are grown for either their seeds or the unripe fruit which can be eaten as a vegetable. Those species which are grown for their dry seed provide protein and/or oil. Many other legumes produce large seed rich in protein and oil and it might be thought that any one of these could be improved to become a new crop species. However, two features mitigate against all legumes. These are hard seed and poisonous principles within the seed. It would appear that improvement of the existing crop species would be more profitable than attempting to upgrade a 'wild' species to crop standards.

The forage legumes

Almost all the small-seeded legumes could be thought of as potential forage crops. The success of any one species will depend on its ability to recover from heavy intensive grazing, or cutting, while its acceptability to animals is determined by its palatability, and lack of poisonous substances. Palatability is not a characteristic of a plant that can be correlated with the morphology, but one feature that detracts from acceptability without necessarily making the plant less palatable is the possession of hairs. Glabrous plants are acceptable to the animal. Unusual flavours can also make a plant less acceptable to an animal.

A few species of herbaceous legumes are creeping and the growing point in these creeping types is situated below the level to which animals can graze, browse or nibble. These species resemble the vegetative grass in this respect and they will always possess the capacity to produce new stem material. However the legumes do not possess intercalary meristems within the leaf so when a portion of a leaf is lost there is no possibility of the production of new photosynthetic tissue to compensate for the loss. Prolonged removal of leaves, or intermittent cutting, with insufficient time between cuttings for recovery, will cause the death of the plant. The management of forage legumes requires greater skills than those needed to maintain a grassland.

Only a very few annuals are grown for forage, but short-lived perennials are often grown mixed with grasses to produce hay. These short-lived species, and also the longer-lived perennials are not often made into silage because the high protein content can result in unwanted fermentations, in which amines are produced. When mixed with another type of plant to give a composition in which the level of protein is reduced then silage making is more successful. Another way in which the composition of the raw silage can be balanced is to add molasses to the legume as it is ensiled. Perennial herbaceous legumes can be grown in pure stand to give a hay crop which can be cut several times in a growing season. When made well this hay is of excellent quality.

The genera which are most important are *Trifolium*, *Medicago* and *Melilotus*, all of the tribe Trifolieae, and *Lotus* of the tribe Loteae.

The clovers, belonging to the genus *Trifolium*, represent the most widespread of the forage legumes. Some 500 species of *Trifolium* are recognized and within the cultivated species there are many forms, some of which have been given taxonomic status. With such a large genus there has been erected a complicated taxonomy. Taubert recognized the subgenera *Trifoliastrum* and *Lagopus* whilst Tutin has accepted as subgenera *Falcatula* and *Lotoidea*. Within each subgenus, regardless of nomenclature, there is a number of sections. What is remarkable is that the cultivated species are grouped in a few sections.

Clovers in temperate agriculture are found mixed with grasses to provide either hay (or silage), or grazing. At one time pure stands of clover were grown for conserving, with sometimes the after-growth being grazed. However, clovers are difficult plants to feed to animals since digestive upsets are easily produced if the animal eats too much clover, or is put to pasture at a time when the plant contains substances which are capable of poisoning it. It has been considered that the cyanogenetic property of the clovers, and most strains and species are cyanogenetic to a degree, was responsible for the digestive upset, but there is evidence for the presence of substances which can paralyse smooth muscle, and if this happens normal peristalsis is prevented. Another phenomenon encountered in animals on a diet rich in clover is infertility. This infertility of females is due to the clovers producing substantial levels of oestrone-like compounds. It is advised that young female animals should not be put to graze on clover-rich pastures if they are intended for breeding purposes shortly afterwards.

The clovers can be divided into two habit types: the erect and the creeping. Within each of these distinct types we find a range of variation so that some of the more procumbent of the ascending species approach the habit of an erect creeping type. The use of the term 'creeping' should be restricted to those clovers which have the capacity to root readily at the nodes. In general the erect forms are not persistent whilst amongst the creeping species very long-lived ecotypes are found. This difference in persistence is due to the way in which the grazing animal depletes the plant of essential photosynthetic tissue and removes terminal meristems. The differences are best illustrated by referring to the two main clovers of British agriculture.

Red clover, *Trifolium pratense* L., is normally managed as a biennial, but when grazing pressure is reduced the plant is a short-lived perennial persisting for from three to four years or more. On germination two cotyledons are produced above ground. These cotyledons are ovate, stalked and glossy. The plumule produces as the first leaf a simple unifoliate leaf, hairy in case of red clover and longer than broad. This leaf has a long petiole. The succeeding leaves are trifoliate, or ternate, the leaflets being obovate, notched, and with a prominent midrib. Like the first leaf the succeeding leaves have long petioles, and the stipules are prominent. The shape of these stipules is of considerable diagnostic value.

During the early stages of growth there is little internode extension, but the rootstock becomes substantial and serves as a storage structure. At flowering the stem elongates by internode extention to reach a height of 75 cm. Inflorescences in red clover are borne in the axils of the upper leaves, and there is no discernible peduncle. The raceme is contracted to a globular head called a *capitulum*. If the above-ground parts of the plant are removed at this time all the active meristems are lost, as are the leaves. Recuperation of the plant consists of activation of such basal axillary buds as are left and the production of adventitious buds on the rootstock. Unlike lucerne (see p. 87), red clover cannot survive more than two such treatments in a growing season, and it soon disappears from mixed vegetation due to the superior regenerative capacity of the grasses. When grazed, its disappearance is even more rapid, although individual plants, especially of 'wild' prostrate types, can often be found in pastures.

In comparison to *T. pratense*, white clover, *T. repens* L., is markedly creeping, especially in those ecotypes found in old pastures. These extreme forms are often called wild white clover. Germination in *T. repens* is similar to that described for *T. pratense* but the seedling is smaller and glabrous, and in white clover the first leaf is broader than it is long. First-formed leaves give rise to a rosette but soon, before flowering commences, the axillary buds become active. The stems that are produced are plagiotropic to varying degrees, and in the prostrate types rooting occurs at the nodes. Some races of white clover are ascendent, and at first the stems do not come into contact with the ground, but as the stems age and lengthen their weight carries them downwards and rooting at the nodes then takes place. The petioles of the trifoliate leaves are long, and when inflorescences are produced they are carried above the leaves on long slender peduncles. The heads in white clover are smaller than in red but they are of the same general form, though made up of white flowers.

A great range of variation is seen in these two species and this variation is associated with the origin of the variant. Three broad groupings are established in red clover:

1. Atlantic
2. Central alpine
3. Continental (meaning Europe)

The Atlantic races constitute the early red clovers, and these can give two cuts of hay in one season. Sometimes they are called broad-leaved cow grasses. These races are not persistent. Continental races are later flowering, and have narrower leaflets than the preceding. Only one cut of hay can be taken from these varieties in a year, and they are thus called 'single-cut cow grass'. Continental races are persistent, but this may be due to management as described earlier, and not to any innate capacity possessed by them but not by the Atlantic races. The central alpine races are intermediate between the other two.

In white clover the situation with regard to the taxonomy of the variants is much more definitive. White clover has a wide distribution from the eastern Mediterranean to Britain. In the eastern part of its distribution the plants tend to be stragglers, and indeed some without creeping stems (stolons) are recognized. Western forms of white clover can be profoundly stoloniferous. The following varieties of *T. repens* are thought to be good taxa: *sylvestre* (the wild white clover), *purpureum* (bloodwort and red flowers), *grandiflorum*, *rubescens*, *biasolettii* and *ochranthum*. Within variety *sylvestre* there are the race *hollandicum* which is the true wild white clover, and race *giganteum* which is the Lodi or Ladino clover of commerce. Ladino white clover, named after a district in Italy, is upright with large flower heads, short peduncles and petioles, but having large leaflets. In all these respects it contrasts with *hollandicum*.

T. pratense belongs to the section *Eulagopus* of the subgenus *Lagopus* while *T. repens* is a member of the section *Euamoria* of the subgenus *Trifoliastrum*. In this section we find *T. hybridum* L. (alsike clover) which is the other important clover of cool temperate regions. In habit it is like red clover but its flowers are pink to white. Indeed it is so intermediate between the clovers described that Linnaeus thought of it as a hybrid. It is not.

We have considered white clover in some detail but our treatment of red clover has not been to the same depth. It was stated that the variation that is found in the agricultural cultivars is correlated with the origin or provenence of the morphological types. Taxonomists have established subspecies in both red and alsike clover, one of which is allocated to the cultivated forms. In *T. pratense* the cultivated types belong to the subspecies *americanum*, so named because these types were derived from a form that was reintroduced to Europe from America in the late 19th century. The other subspecies *sativum*, a short lived perennial, *maritimum* and *frigidum* are all considered to be 'wild' although they are likely to have been progenitors of the forms now in cultivation.

With alsike clover much the same taxonomic treatment has been adopted and three subspecies are recognized. Subspecies *hybridum* includes all the cultivated types while the 'wild' races can be allocated to either of the subspecies *elegans* or *anatolicum*.

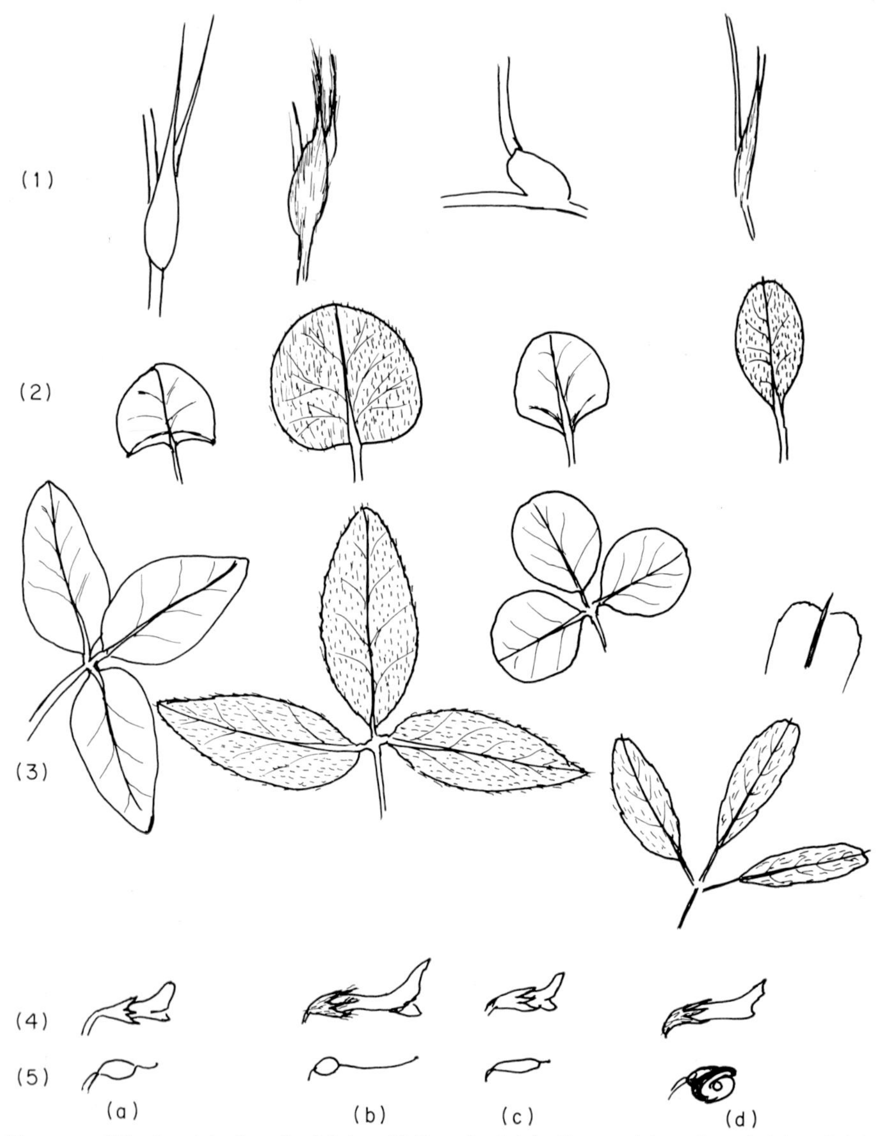

Fig. 24. Stipules (1), first leaf (2), trifoliate leaf (3), flower (4) and pod (5); of (a) *Trifolium hvbridum* (b) *T. pratense* (c) *T. repens* and (d) *Medicago sativa*. Note the mucronate tip to the leaflets of (d).

In the warmer regions where it is still possible to grow forage clovers the most important species are *T. subterraneum* L. (subterranean clover) which is an annual with a habit similar to ground nut in that the peduncles of the ripening flower head elongate to take the infructescence to ground level or even bury it, and *T. alexandrinum* L., berseem. Subterranean clover is native to north-west Europe but in its native habitats it is not productive. When introduced to Australia it proved to be one of the most successful of the forage legumes. Berseem is native to North Africa, and in its original localities it is one of the best plants for the production of fodder. It responds particularly well to irrigation.

The outstanding forage legume of the warm temperate and subtropical regions of the world is lucerne or alfalfa. It is grown as a hay crop, or in the subtropics it provides material for zero grazing. The adult plant will grow well in cooler regions but the difficulties associated with establishment and the possible high incidence of crown wart caused by *Urophlyctis alfalfae* (Lagerh.) Magnus, have resulted in its not being grown in the wetter, cooler regions of Europe and North America.

Lucerne, *Medicago sativa* L., is a long-lived perennial which develops a robust root stock and crown. The root system is massive and is stated to penetrate to depths of 3 metres. The crop is able to thrive on relatively dry soils and this ability is ascribed to the deep root system. However, the highest yields are only obtained when the crop is plentifully supplied with water.

In habit the plant closely resembles red clover and like it lucerne is softly or markedly hairy. There are distinctive features which separate the genus *Medicago* from *Trifolium*. The seeds are alike in size but the seed of lucerne is narrower. Germination is epigeal and the cotyledons are strap-shaped without a separate stalk. As with the clovers the first leaf is unifoliate, but the blade is deeply toothed, with the midrib projecting beyond the edge of the lamina. This prominent terminal projection is called a *mucronate tip*. The stipules are narrow and pointed. The adult leaves are trifoliate and each of the leaflets is narrow, serrate, and with a prominent mucronate tooth. Furthermore the central leaflet is distant from the laterals. Some species of clover have leaves similar to this, but they are not among the cultivated species.

The seedling does not develop as a rosette, internode elongation occurring as leaves are produced. The adult plant can have stems up to 1 m in length and is much branched. In old plants the branching is partly from adventitious buds on the crown. The inflorescences, which are axillary, are many-flowered and moderately contracted, and in the cultivated forms the flowers are blue through to pale lilac, though yellow-flowered types are found. The fruits are coiled, containing several seeds. Cross-pollination is required and the flower possesses a trip mechanism causing it to explode with a single visitation.

Species of forage legume which were more widely grown in the past, or grown locally because disease prevented exploitation of the species described, or tropical species which are presently under examination as potential crops are listed below. The successful species will have the habit of either white clover or lucerne depending on whether it is grazed or cut, and providing it is as productive, palatable and easy to manage as these it may oust them in a particular region if they are outyielded. Since the beginning of scientific agriculture new 'rivals' to the forage legumes discussed have been found, but in critical assessment trials the established species have proved equal or superior to the 'rivals'. The most widely grown species are as follows:

Trifolium incarnatum L., crimson clover: an annual not subject to the same diseases and pests as red, alsike and white clover.
T. dubium Sibth., an annual, with yellow flowers; exhibiting much variation and now considered as an aggregate species.
Medicago lupulina L., black medick, hop trefoil: annual or biennial and used like crimson clover where there had been a build up of clover diseases and pests. The build-up of these resulted in 'clover sick' land.
Onobrychis sativa Lam., sainfoin: grown instead of lucerne.
Ornithopus sativus Brot., seradella: an annual suited to dry soils.
Anthyllis vulneraria L., kidney vetch: a rhizomatous perennial which can withstand grazing pressure.
Lotus corniculatus L., bird's foot trefoil.
L. pedunculatus Cov., greater bird's foot trefoil.
Melilotus alba Desr., white sweet clover, Bokhara clover.
Stizolobium Hasojoo Piper and Tracy, yokohama bean.
S. duringianum Bort., Florida velvet bean.
Stylosanthes humilis H. B. and K., Tounsville stylo.
St. Guyanensis Sw.

The taxonomic treatment we have applied to the Lotoideae has been based on that given by Taubert in Engler and Prantl. More recent treatments have expanded the number of tribes, but the essential affinities of the crop species are evident in the older classification. If our treatment of the crops had been based on one of the newer classifications the artificiality of the tribe in the Lotoideae would have become apparent since the key is equivocal. Taubert's tribes were based on the ten erected by Bentham, but Hutchinson in his consideration of the Lotoideae establishes 50 tribes. We consider that such a treatment is not necessary to appreciate the crop legumes.

CHAPTER 3

The Solanaceae: Potato, Tobacco and Allies

The Solanaceae is a family which has its main centres of distribution and variation in the Americas. Only part of one tribe is represented in the floras of the Old World. This has meant that the members of the family which have been exploited as crops have only been in the agriculture of Europe and Asia since the 15th century, though some of the species have been cultivated in the Western Hemisphere ever since agriculture was developed there, perhaps some 10 000 years ago.

Compared to the grasses and the legumes the potato family is quite small, having some 75 genera and 2000 species. However, two of the genera account for well over 1000 of the species; indeed the single genus *Solanum* has at least 1000 species, but as we shall see the interpretation of the species in a part of this genus might be different from that usually adopted.

Most of the members of this family are herbaceous, but shrubs and small trees occur. Regardless of their stature, Solanaceous species are readily recognized. Nearly all have hispid hairs and are strong-smelling, the smell becoming more obvious if the plant is bruised. A few are glabrous, but these are exceptional and the strong smell of the family is manifest when the leaf tissue is broken. It is not the smell alone that conveys upon these plants their most distinctive feature; this is seen in the way in which the leaf and the stem are associated.

The leaves may be simple or compound, the latter having a pinnate pattern to the distribution of the leaflets, but these are not all alike, nor are they arranged in distinct opposite pairs. The form of the lamina will be considered later in detail under the description of the potato. A petiole is present but stipules are absent. Immediately below the apparent point of insertion of the leaf on the stem it is seen that the stem carries two ridges running down from the edges of the petiole. These ridges may be prominent and so well developed that the stem is winged. The ridges terminate some distance below the point where the petiole springs from the stem, and it can be seen that the petiole is in fact joined to the stem along the lower part of its length and the ridge is in fact the junction between these two structures. When two dissimilar organs are joined together they are said to be *adnate* and adnation of the petiole and stem is a distinctive feature of the Solanaceae (Fig. 25). Anatomical studies confirm the adnation. The true node is situated on the axis at the point where the vascular system branches to give the leaf and bud traces, and this is seen externally as the point of origin of a stem ridge. The axillary bud is carried up from the node and is found where the petiole springs from the stem.

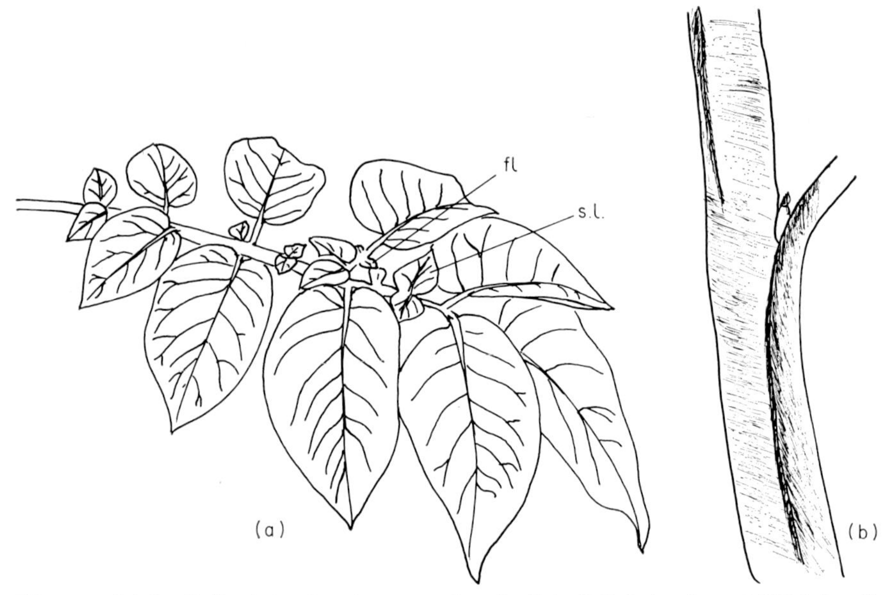

Fig. 25. (a) Leaf of potato, showing secondary leaflets (folioles), s.l., and foliolules, fl, and the adnation of the base of the petiole to the stem (b).

In vegetative stems the adnation does not lead to complications in the arrangements of leaves, which are alternate, but in the region of the axis carrying inflorescences the adnation between petiole and stem is coupled with adnation between peduncle and stem, and extremely complicated arrangements of leaves and inflorescences can occur. In many species it is observed that two leaves, a small one and a large one, appear to arise together at a point which carries a flower or an inflorescence. Other arrangements result in the inflorescence appearing to originate as a branch from the axis, and not be associated with a subtending leaf. Each case must be treated and interpreted separately.

The flowers are carried in determinate, or cymose inflorescences, though solitary flowers are found. In some species the cymes are developed in such a way that they are superficially like racemes (an overtopped cyme). This is the case in tobacco. The individual flowers are showy, and some species have been selected as ornamentals.

The flower in this family is tubular, the calyx members being fused at their bases, and likewise with the petals, where the lower parts fuse to form a distinct corolla tube. The terminal portions of the petals may or may not be distinct, and if this upper part of the petal is narrow the flower is termed *stellate*. When the terminal part of the petal is not noticeably prominent the flower is considered to be *rotate* and these two extreme types of corolla can be found within a genus. Both the calyx and corolla have five members and in most of the genera they are symmetrical, though the genera within one of the tribes have corollas which are zygomorphic.

Normally the stamens, again five in number, are epipetalous and the anthers are connate. Pollen is shed either by the anther splitting along the outer sutures, or through apical pores. Occasionally there is abortion of stamens, seen especially in those species with zygomorphic corollas.

The gynoecium is superior and consists of two carpels fused together but they are positioned such that they are asymmetrical in relation to the axis of the flower. Because of this the gynoecium is zygomorphic. The ovary is bilocular with a thick fleshy placenta on which are carried many ovules. The placentation is axile. In a few genera ingrowths from the carpellary wall divide the primary loculae further and the mature ovary in these cases is multilocular, though quite regular. In many of the cultivars of tomato there is a degree of proliferation of floral parts, and also fasciation within the developing inflorescence. This leads to flowers with more than five members to each of the whorls and to fruits that are multilocular, but not regular. There is a single style with a globose stigma. At anthesis the stigma may or may not be extruded. Fruits are berries or capsules.

The seeds are endospermic and the embryo is curved or spiral lying within the endosperm. In general the seeds are small.

This family is one in which there has developed a range of toxic alkaloids and saponins. Some of our most poisonous plants belong to the Solanaceae, yet so does the most important of the world's starch crops, the potato. The only other crop which gives a comparable yield of starch directly consumable by man is banana, but since the acreage of potatoes is many times that of banana, the total production of the former far exceeds that of the latter. In all probability a greater weight of potatoes is produced than of any other crop, but it should be remembered that potatoes are only about 20% dry weight, whereas the cereals are at least 87% dry matter.

Many Solanaceous plants were grown for medicinal purposes and are still to be found around the sites of old castles and monasteries. The drugs atropine, hyascine and belladonna were and are derived from members of the Solanaceae. Usually all green parts of the plant are poisonous, though in a few species this is not the case, especially of fruits. However, it would be unwise to rely on the premise that a non-green part of a plant is safe to eat. In the genus *Solanum* some species have edible fruits and others have poisonous berries, and even within a species, e.g. *Solanum nigrum* L., black nightshade, there are varieties with poisonous fruits and others with non-poisonous berries.

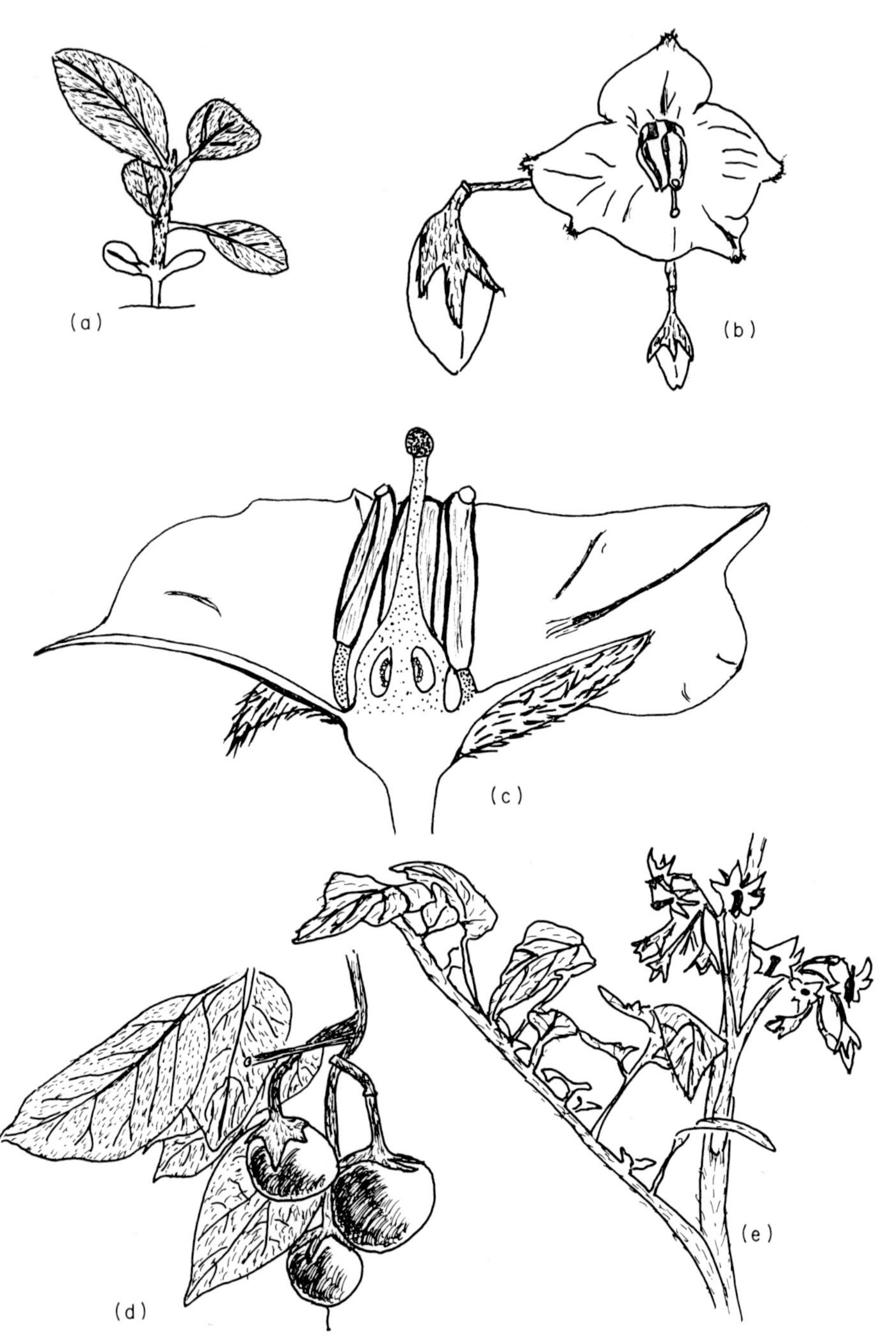

Fig. 26. Potato. (a) Seedling. (b) Flowers. (c) Half-flower. (d) Fruits. (e) Inflorescence of tomato.

The taxonomy of the family is straightforward but when the taxonomy of the genus *Solanum* is considered it is found to present some difficulty, especially in deciding on the limits of a species. In von Wettstein's treatment of the family five tribes are recognized, and these are separated on the basis of embryo form, the ovary, and whether the stamens are all fertile. A practical division is as follows:

Nicandreae:	Spiral embryo, ovary wall ingrowths resulting in 3–5-celled fruit.
Solaneae:	Spiral embryo, 2-celled fruit, no ingrowths.
Datureae:	Spiral embryo, 4-celled fruit, one ingrowth into each primary loculus.
Cestreae:	Straight or curved embryo, all five stamens fertile.
Salpiglossideae:	Straight or curved embryo, some stamens sterile.

The economically important genera belong to the tribes Solaneae and Cestreae, though ornamentals have been selected from all the tribes of this family. The most intesting of all the genera is *Solanum* itself.

The Genus *Solanum*

As remarked, this is a very large genus with perhaps more than 1200 species. *Solanum* is distinguished by having anthers which dehisce by pores, though in old anthers the pore may be extended by splitting of the back of the anther, but this should not be confused with normal extrorse dehiscence. The fruit is a berry.

This large genus has given us the potato and the egg plant.

The genus is divided into six subgenera, and the subgenus *Solanum* is composed of 25 sections, one of which is *Tuberarium*. *Tuberarium* is in turn divided into the subsections *Basarthrum*, in which the articulation of the pedicel is at its base, and *Hyperbasarthrum* where the articulation of the pedicel is above the base. Other characters are correlated with the position of the point of articulation. In *Basarthrum* the plants are covered in bicellular pointed hairs, are probably without tubers, are straggling and perennial. Plants belonging to the subsection *Hyperbasarthrum* have multicellular hairs, and are usually tuber-bearing, upright, with annual stems. Furthermore plants of the first subsection have an unpleasant foetid smell.

The taxonomy of the subsection *Hyperbasarthrum* has been the subject of much investigation especially since the economic importance of the potato has been responsible for encouraging collecting trips to the centre of variation. This centre is in the region extending from New Mexico and Arizona down through

Central America to Argentina, keeping mainly to the higher altitudes of the Cordilleras. In this region seeds are produced with regularity and the species are cross-pollinated. Species hybrids do occur. This centre of diversity is probably also the centre of origin of the genus and within it the *Solanums* are plants of regions of ecological instability. In the stable communities it appears that the pioneering *Solanums* are displaced in favour of the plants of the climax vegetation. The possession of tubers which might be thought of as an efficient perennating system does not seem to confer upon the plants an ability to survive for long periods in competition with other species. It is usual to find that species which are typical of disturbed habitats and which have an efficient outbreeding mechanism, leading to possible interspecific hybridization, exhibit a great deal of variation. The exact nature of the genetic affinities of the various individuals can often only be analysed by extensive genetic studies in the greenhouse and experimental garden.

When grown away from their area of natural distribution many of the *Solanums* do not show the same type of breeding system. Perhaps this is better seen in the closely related tomato *Lycopersicum esculentum* Mill. where in America it is outbreeding, but in Europe, especially towards the northern parts, it is inbreeding. This is because in the warmer areas of short days the stigmatic surface is exerted by the style growing beyond the rim of the anther tube and allowing pollinating insects to carry pollen to it. In the cooler long days of Europe the style does not elongate to the same extent with the result that pollen is shed into the anther tube when the flower is disturbed, leading to self-pollination. Something similar happens in *Solanum* but the degree to which it takes place is species dependent.

Not all of the members of the subsection *Hyperbasarthrum* have provided man with potatoes. Outside the area of its natural distribution only one species, *S. tuberosum* L., is grown for its tubers, but in the Americas the South American Indians utilize the tubers of a number of species. The improvers of *S. tuberosum* have looked upon these other species as being of possible use in their breeding programmes, and indeed genes from the 'wild' species can be incorporated into the 'domestic' potato.

The result of the very many studies of the subsection *Hyperbasarthrum* is that it is now divided into a number of series. There are two main methods of division in use today, one of which establishes 17 series, and the other twenty-six. In either case the series *Tuberosa* is the one that is of greatest concern to us.

Within most of the subsections it is usual to find a polyploid series, and in this respect *Tuberosa* is quite typical. Most of the species are diploid, some are tetraploid, and triploids and pentaploids occur. Both the triploids and pentaploids must represent the result of hybridization or the fusion of unreduced gametes. Hawkes lists seven species of this subsection he considers to be cultivated types, but this list does not include all of the potato species which are exploited in the South American continent. Of the seven cultivated species only one, *Solanum tuberosum* L., is grown on a worldwide scale. The origin of this plant has been the subject of considerable controversey but it is generally agreed that it has arisen in culture, probably in response to long days and blight infestation. The present-day potato does not correspond to the plant that is illustrated in the literature of the 16th century. A plant which agrees more with these illustrations is *S. andigena* Juz. and Buk. and it is probable that this type was ancestral to *S. tuberosum*. It is best to consider that *S. tuberosum* has two subspecies, *tuberosum* and *andigena*. In all probability *tuberosum* has arisen more than once, at some time in the past in Chile and again in Europe. Recently selections of seedlings grown in the United Kingdom under the extreme blight conditions encountered in the wet south-west of the country have given rise to a plant that has been called *neo-tuberosum*. This particular exercise was carried out to enlarge the genetic base from which plant breeders could select material for incorporation into commercial varieties.

Solanum tuberosum subsp. *tuberosum*. The Potato

The potato is normally propagated by stem tubers but reproduction by seed can take place. Our description of the plant will begin with an account of seed germination and seedling development.

Seeds of potato are small (about 1–1.5 mm long) slightly flattened and with one side asymmetrically concave. It is covered with floccose hairs. Germination is rapid, though some samples can exhibit dormancy, and on germinating the cotyledons are brought above ground. These cotyledons are stalked and the ovate blade comes to an abrupt point.

The first-formed leaves are simple and noticeably hairy. As the plant ages the leaves that are produced become compound, there being a series of leaflets arranged pinnately along the distant region of the leaf stalk. The basal part of the leaf stalk is adnate to the stem, but the central region free of leaflets and not associated with the stem is distinct and is talked of as the petiole. It should be emphasized that it is only part of that structure. The leaflets are stalked, these stalks being called petiolules. Between the points of insertion of the petiolules pieces of lamina are sessile on the main leaf stalk. These structures have been called folioles, but Hawkes, amongst others, prefers to call them secondary leaflets. Similar small pieces of lamina can be found on the petiolules and these are termed foliolules. The pattern of lamina development can be a very useful diagnostic aid towards identifying both species of *Solanum* other than *tuberosum* and cultivars of that species. Generally the margins of the lamina pieces are smooth and shallowly scalloped. A mutant form with coarsely dentate margins exists, the so-called tomato-leaved form. (In tomato there is the potato-leaved mutant with a lamina like that described.)

The cotyledonary buds of the young seedling grow out horizontally as stolons with long internodes. The leaves at the nodes along the stolon are vestigial. Soon the stolons arch downwards to bury the tip in the soil. The basal buds immediately adjacent to the cotyledonary node behave in a manner similar to the cotyledonary buds. The young seedling then will have some two to five stolons, each with the distal portion underground. Below ground, axillary buds on the stolon may become active leading to branching of these organs. The tips of the stolons begin to swell radially without the internodes elongating as they had done in the proximal region so giving rise to a tuber. Stolon development can be weak or strong, and in commercial cultivars the stolons are not extensive.

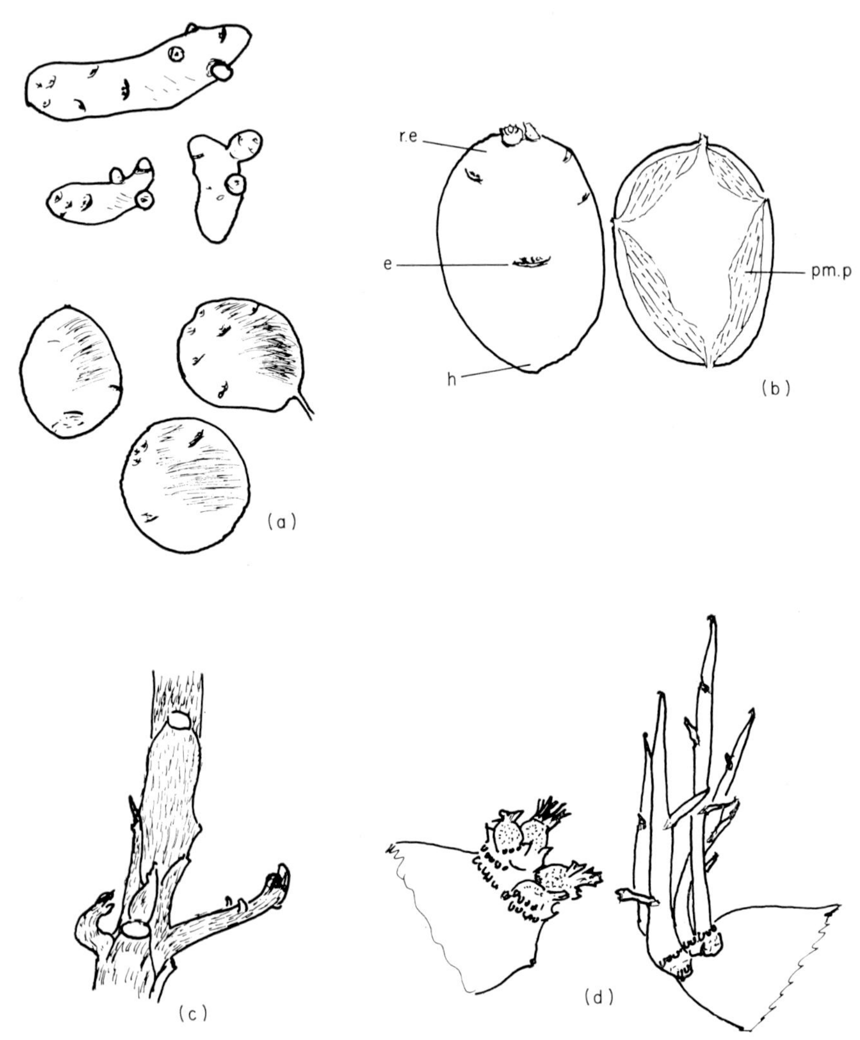

Fig. 27. Potato. (a) Tubers of ssp. *andigena* (top) and ssp. *tuberosum* (bottom). (b) Tuber showing eyes, and sectioned to indicate the extent of the proliferation of the perimedullary phloem (pm.p), rose end (r.e.), heel (h), eye (e). (c) Axillary branches developing as stolons when plants grown under fluorescent lamps. (d) Sprouting of tubers in light (left), or darkness (right).

The tuber that is produced in potato is the result of an anomalous type of thickening not seen in any other family of flowering plants. The family Solanaceae has a stem anatomy which is uncommon. Firstly, in the primary condition there are seen in a section through an internode not only the main vascular bundles of the stem, but also the traces to the leaf and the bud. This is because of the adnation of the petiole to the stem. When secondary thickening takes place the individual bundles lose their identity but evidence of them can still be seen in the primary xylem. Secondary thickening in the stem is normal.

As well as the bundles showing dimorphy, there is found a well-developed perimedullary phloem (Fig. 28). Islands of sieve tubes and companion cells are present, concentrated close to the protoxylem and also with isolated groups between the bundles. Each group of perimedullary phloem has associated with it its own parenchymatous tissue, which is made up of cells smaller and more rounded than the parenchyma making up the pith proper. In the stem and the stolon, the pith parenchyma may disintegrate to give rise to a pith cavity, but the parenchyma of the perimedullary phloem remains. In the tuber both the pith parenchyma and the perimedullary phloem parenchyma proliferate and the resulting cells become filled with starch. However, the cells which are formed by the proliferation are dissimilar and the regions which contribute to the tuber can be seen quite distinctly. The central medullary cells are large and loosely filled with starch, but the cells derived from phloem parenchyma are smaller and densely packed with starch. On cutting a tuber, transversely or longitudinally, the central region has a 'cheesey' appearance, while the outer zones look 'chalky'. The extents of these two regions are dependent on the cultivar in question, and the use to which the potato can be put may be determined by the relative amounts and character of pith and phloem derived tissues in the tuber.

After stolon formation the axis branches in the upper region and on these upper branches inflorescences are produced. These are many-flowered cymes with white, lilac, or purple flowers. The pedicel is articulated in the middle third, and the flower has short calyx lobes. The corolla is rotate, or stellate rotate. In other respects the flower agrees with the description given for the family. The fruit is a bilocular berry which at maturity is purple, tinged with green.

For crop purposes the tuber, not the seed, is used to plant the crop. The end of the tuber attached to the stolon is called the *heel end*, while the apex is termed the *rose*. Axillary buds, the *eyes*, are distributed unevenly along the axis of the tuber, most being concentrated towards the rose end. An *eyebrow* subtends each eye. This eyebrow is in fact the scale leaf that is associated with the buds of the eye. As harvested the tuber in this species is dormant but the dormancy wears off in storage. When dormancy is lost the buds in the eyes nearest the rose begin to grow.

If the tuber is planted before the buds in the eyes grow out, the sprout which forms has long internodes with a small number of subterranean nodes. At each node there arises a ring of adventitious roots and the sprouts soon become

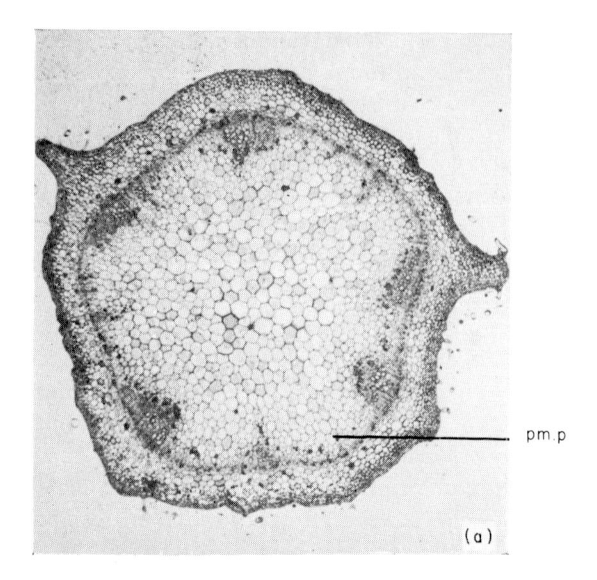

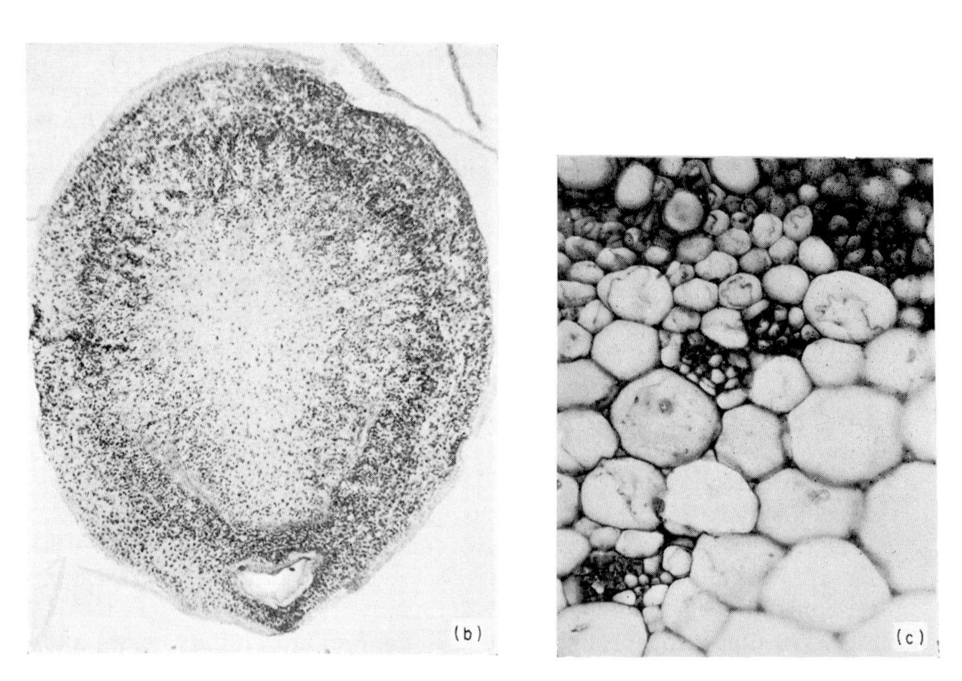

Fig. 28. (a) T.S. of young potato internode: pm.p., perimedullary phloem. (b) T.S. of young developing tuber. (c) High-power view of islands of perimedullary phloem.

independent of the mother tuber. When the sprouts come above ground they turn green. While underground the sprouts have a hooked apex similar to that found in the case of plumular shoots. Any leaves which are present below ground are not well developed and may be present as no more than a scale. The aerial shoots produce leaves of the adult form. From the subterranean nodes shoots arise, and grow plagiotropically. Though these shoots are subterranean they are called stolons. These stolons may branch, so building up a considerable below-ground axis complex. Eventually growth and proliferation of stolons slows, and this slowing of the growth which led to the amplification of the stolon system is accompanied by swelling of the stolon tips to produce tubers. It can be seen that the degree of ramification of the stolon system is primarily dependent on the number of subterranean nodes on the sprouts.

The total weight of tubers that can be produced in any given set of circumstances could be made up of either small or large tubers. If the tubers are small there will be many, and if large, few. It is desirable to have large tubers if the crop is required for human consumption (a ware crop), and the regulation of stolon numbers per unit area of land is achieved by the following morphological features: the number of shoots per unit area of land; the number of subterranean nodes per shoot; the degree of stolon branching.

The first of these characters is influenced by the number of shoots produced by each tuber planted and the density of planting. To obtain few shoots per tuber it is necessary to plant small potatoes. To reduce the density the tubers are planted widely apart in rows that are themselves widely spaced. In the production of a ware crop, that is one grown for human consumption, the practice is to plant small tubers (seed potatoes) at a moderate density (about 100 000 per hectare).

It is possible to manage the number of nodes below ground by sprouting the tubers in the light. This suppresses internode extension, so ensuring that a greater number of nodes are present at the time the tuber is planted. The practice of sprouting potato tubers in light is called *chitting*. Sprouts which develop in light are robust, have short internodes and do not have the hooked tip. When planted there are many subterranean nodes from which stolons can grow, leading to the possibility of a large number of tubers from each sprout. As a result there is a tendency for each tuber to be smaller. This is the type of tuber that the grower of 'seed' potatoes desires and it is usual that for 'seed' production sprouted tubers are planted. Sprouting potatoes ahead of planting can result in advancing the emergence of the crop. In later districts this can be an advantage.

There is no practical way of controlling stolon branching.

Branching of the potato shoot takes place not at the base but in the upper zone. The complicated pattern of branching is associated with the production of inflorescences, but it is not simply a morphological development, since it has been shown that the nature of the shoot produced by an axillary bud is controlled by the duration and quality of the light that is incident on the plant. It is this photomorphogenetic phenomenon which gives rise to plants of very different form when members of the same clone are grown in different environments. Indeed it was this which led to the many and varied descriptions of potato shortly after it was introduced to Europe, and which today gives rise to controversy when potato species are described from their native habitats, and from cloned material grown elsewhere. Perhaps after more is known of the physiology and genetics of the subsection *Tuberosa* we shall have a better idea of the nature of a species in this most important taxon.

The only other member of the genus *Solanum* which has been taken into culture is *S. melongena* L., the egg plant or aubergine. Egg plant is grown for its very large fruit, which can exceed 20 cm in length and be 10 cm in diameter. It is grown throughout the tropics, and in the United Kingdom with the large immigrant population demanding their usual vegetables a considerable production under glass has been established. Its culture in temperate climates requires similar conditions to those used in the production of tomato.

In its natural habitat the plant is perennial but in culture it is grown as an annual. The seed is small and germination is epigeal. The seedling soon develops an upright axis, but as branching becomes significant the plant becomes straggling. The stem and leaves are hairy and in old plant robust spines are produced. The inflorescences appear to be extra-axile but this is as a result of a complicated sympodial growth pattern which will be described in detail under tomato (see p. 101). These inflorescences are few-flowered, even reducing to single flowers. The flowers are typical. Only one fruit forms per inflorescence, and not every inflorescence is fructiferous. The most popular cultivars have fruits which are a deep rich purply-black, but fruit of other colours can be found.

The tomato, *Lycopersicum esculentum* Mill., at one time *Solanum lycopersicum* L., is however the most important member of the family grown for its fruit. Colloquially it is called a vegetable, but this is only because of its culinary uses. Like potato, the tomato is a plant of the Americas. Both the potato and the tomato are intolerant of low temperatures, and in the cooler parts of the region where the tomato is grown it must be kept under glass. While not photoperiodically sensitive with regard to the induction of flowers, the plant cannot tolerate long days, unless there is a sharp differential between day and night temperatures. For some unknown reason the photosynthetic apparatus in this plant can only function at maximum efficiency if the chloroplasts are given a dark rest period which exceeds eight hours. If, instead of a dark rest period the plant is exposed to a 'night' temperature some 8 °C less than the day temperature, the chloroplasts seem to be able to remain as efficient photosynthetic units even though the daylength exceeds 16 hours.

The seed of tomato is 2.5 mm in diameter, and is approximately disc-shaped. In *L. esculentum* the seed is covered with long hairs but in the related *L. pimpinellifolium* (Jusl.) Mill. the seeds are hairless. This latter species is the currant tomato, and while not as extensively grown as the ordinary tomato, in some areas it is an important crop.

Germination is epigeal and the cotyledons are strap-shaped, with an abrupt point. The hypocotyl is at first hooked, and is it straightens the cotyledons are pulled from the seed coat, to expand. Sometimes the cotyledons are not fully withdrawn from the testa and the tips are held together. Seedlings with this characteristic never grow as fast as those in which the cotyledons separate and adopt a horizontal attitude. The plumular bud at this stage is small. This bud enlarges and produces an axis which is clothed in robust hairs, some of which are glandular. Compound hairy leaves in which both stalked leaflets and secondary leaflets occur are produced, sometimes, in the case of the older leaves, a bipinnate development of leaflets taking place. The leaflets are coarsely dentate. When bruised the plant has a characteristic odour. A minimum of ten leaves are produced and then an inflorescence is formed. Those cultivars which become floriferous after ten nodes are classified as early but there is a range of cultivars in which inflorescences are not produced until after 13, 16, 19 or 22 nodes. This inflorescence is terminal and is a monochasial cyme, although occasionally branched cymes are formed. However, because the peduncle is adnate to the petiole of the uppermost leaf, the inflorescence is displaced sideways and the axillary bud at the uppermost node grows out to give the appearance of continuing the axis. The penultimate axillary bud grows out, but less strongly, and the development of buds proceeds basipetally, quite the opposite from that expected if a strong apical dominance had been removed. This pattern of growth is like that of potato where we said branching is prominent in the upper part of the plant.

The strongest axillary shoot produces three nodes and then a terminal inflorescence, with again the peduncle adnate to the petiole of the uppermost leaf. The branching pattern is repeated: three more leaves, an inflorescence, and so on. If all the axillary shoots are allowed to grow the plant becomes a straggling bush, but in protected cultivation the 'side shoots' are removed and the plant is trained as a single 'stemmed' vine. When the latter, a mature plant can carry up to 20 inflorescences or trusses, and yields exceeding 250 tonnes per hectare are obtainable. When grown outdoors as a bush yields seldom exceed 50 tonnes per hectare.

The flowers are usually aberrant, for instead of having a five-membered calyx and corolla, the structures have more than five sepals and petals respectively. The petals are yellow and the corolla stellate. Even an individual can have flowers with differing numbers of floral parts. This has arisen by selection. Also the stamens may be unusual, and as remarked the ovary is multilocular, not bilocular. In the production of fruit the viability of the pollen is important and at high latitudes much of the pollen can be sterile. Also at high latitudes the stigma is retained within the staminal column. These two features serve to reduce fruit production, and should be taken into account when the tomato is grown at the limits of its range.

Another genus within the tribe Solaneae is *Capsicum*. The fruits of these, the peppers, are eaten raw in salads, or cooked along with meat, and are not to be confused with the condiment pepper which is obtained from the fruits of *Piper nigrum* L. belonging to the family Piperaceae.

The taxonomy of the genus *Capsicum* is confused, but two species are distinct. An annual form which carries single fruits at each point where flowers were borne, *C. annuum* L., and a perennial which carries its fruits in groups, *C. frutescens* L. The confusion arises at the subspecific level for there are many fruit shapes and colours and these have been used by some authors as the criteria for the erection of subspecies. Since there is free exchange of the genes responsible for these characters it is hardly justifiable to consider them to be of value in delineating subspecies.

In both peppers the seedling stages are similar, a small seedling with two cotyledons of the same general shape as those produced in *Solanum* being the product of germination. The plumule soon expands and there develops an upright axis which produces in its upper region branches and inflorescences. The leaves on the axis are entire and glossy. The leaf arrangement in the vegetative parts of the axis is spiral, but in the inflorescence bearing part a complex adnation pattern results in leaves often appearing to arise as pairs from a common 'node'. The leaves do not have their origin at a common node but because of differing extents of adnation their petioles leave the axis at the same place.

The remaining member of the Solanaceae, which we shall consider is tobacco, *Nicotiana tabacum* L. belonging to the tribe Cestreae. Once again this is a plant which originated in the Americas. It and the closely related *N. rustica* L. were the tobaccos introduced to Europe shortly after the original colonization of the Western World. While *N. rustica* is still grown in some parts of the world, its culture is predominantly for the extraction of nicotine for industrial use, or the preparation of the leaf for insecticidal purposes. A small amount of *rustica* leaf is used in tobaccos that are chewed or smoked. The main tobacco is now *N. tabacum*, which is not found wild.

Tobacco is a large herb usually grown with a single stem but many-stemmed plants can occur. Initially a rosette of obovate stalkless leaves is produced but when the plant is induced to flower, a robust stem arises. Simple ovate lanceolate or elliptical leaves are arranged spirally on the stem and the adnate petiole is manifest by the small wings along the length of the stem. The midrib of the leaf is prominent, and at the base of the lamina there is a pair of auricles clasping, but not encircling the stem.

The whole plant is covered in glandular hairs. The inflorescence is carried terminally and is a dichasial cyme, but soon one set of branches predominates and the cyme is asymmetrical. Because of the way in which the peduncles grow the inflorescence looks superficially like a raceme. Smaller inflorescences can arise from the upper axillary branches.

The flowers have a tubular calyx, and the corolla has a long basal tube terminating in five acute lobes. The five epipetalous stamens have long filaments, some overarching the stigma. In the gynoecium there are two fused carpels and the fruit at maturity is a bilocular capsule. The style is long and surmounted by a globose sticky stigma.

Tobacco produces an abundance of very small seeds, each one less than 1 mm in diameter. It has been estimated that one large plant can produce over 1 000 000 seeds. Germination is slow and is subject to control both by light and temperature, but the source of the seed determines the exact nature of the control mechanism involved in germination, some batches not exhibiting any signs of delayed germination, while others possess a very deep dormancy. On germination two small cotyledons are brought above ground and the first leaf is just over 1 cm long and circular. Initially the seedling is a very slow grower but with each succeeding leaf the growth becomes more rapid.

The production of tobacco is not completed with the growing and harvesting of the crop. When a crop is grown for large thin leaves which can be used as cigar wrappers the plant is grown under shade. In certain areas the shade is provided by sheets of muslin set on high poles to provide a canopy which reduces the level of sunshine to about half that in the open. In this case the management of the growth of the crop determines in part the product, but with this and other tobaccos the post-harvesting procedures are responsible for producing a given type of tobacco.

To ensure the best quality of leaf the inflorescence is removed, as are any secondary branches that arise as a result of removal of the tops. The leaves are harvested singly or the whole plant is cut at a time when the maximum number of the large basal stem leaves have reached a stage prior to the onset of advanced senescence. Few tobaccos are produced from leaves fermented before curing, though present-day methods of curing leaves allows enzymatic action to proceed during the early stages of wilting and yellowing of the leaf. Different rates of curing are possible, and the commonest are air curing, flue curing, fire curing and sun curing. The last is the least predictable.

After curing the leaves are placed in heaps, and the moisture content is such that some fermentation is possible. With coarse material the midribs may be removed at this stage.

The types of tobacco that are on the commercial market have arisen as much from variations in curing and bulking as from morphological and genetic differences encountered within the species. This is the case with a great many plantation crops where the item which enters trade is plant material subject to an on-farm process prior to sale.

The nicotine in the leaf is the drug which satisfies the smoker and it is strange that in the Solanaceae which is so abundantly provided with alkaloid-containing plants only one, tobacco, is widely grown for its active principle.

CHAPTER 4

The Palmae: The Palms

Palms have played, and continue to play, an important part in the life of man living in the tropics. Plants from this family can provide man with all his requirements, for food, material for clothing and housing, and stimulants.

Among plants the Palmae are unique. They are monocotyledons and in this class the tree habit is rare, being seen only in the palms and a few tree species closely related to the Liliaceae. But in the palms the tree habit is the result of primary growth, whereas with the other trees secondary thickening has been involved in building up the plant body.

Palms are found within the tropics, though a few species can be grown outdoors in the temperate region, with possibly those found in the sheltered west coast of Scotland being those growing furthest from the equator. Fossil evidence shows that the palm was much more widespread in the past than it is today. The palms of the New World are considered distinct from those of the Old, and it seems as though only a few genera have representatives in both geographic regions. By the standards we have adopted, the family is of moderate size, having some 236 genera and 3400 species. Since the plants are large, the collection of material is difficult and the great herbaria of the world cannot keep specimens for examination by taxonomic experts. To appreciate the taxonomy of this family it is necessary to examine the plants as they grow, and few botanists have had this privilege. In temperate regions, only where there are large botanic gardens is it possible for collections of relatively small numbers of species to be established.

Palms are perennials which take a long time to establish themselves, remaining small for many years then producing a stem which can reach 30 m or more. The early period of growth is associated with the lateral growth of the axis, and once there has been built up a base on which the tall stem can rest, growth in height takes place. This juvenile stage, during which the axis remains unextended, may last as long as 20 years. Since all the growth is primary, there is no anatomical evidence that permits us to gauge the age of a plant as can be done with trees that produce seasonal growth rings, and estimates of the longevity of palms must be based on trees that have a documented history of germination or transplanting when juvenile. Some of the plants in the gardens of the diplomatic quarters of the former colonial governments do have this type of documentation, and as these old plants die we may be on more certain ground when estimating how long palms live. It seems that they do not live much beyond 200 years, and in the case of some monocarpic species the lifespan is much less.

Germination takes some time, in the case of date over a month at a constant 30 °C. The embryos are small in relation to the size of the seed and the cotyledon elongates to push the radicle and plumule, which is submerged in cotyledonary tissue, away from the seed through the operculum. The cotlyedon tip remains embedded in the massive endosperm, and functions as a haustorium. The radicle elongates rapidly, and penetrates the soil to varying depths depending on the species.

A few lateral roots are formed on the radicle, but the main root system develops on the hypocotyl, which is beginning to enlarge laterally, and become a substantial structure in the shape of an inverted dome. At about the time that the radicle starts to produce lateral roots the plumule bursts through the base of the cotyledon and the first leaf, which is lanceolate and entire, comes above ground. The second leaf may have two pinnae, though sometimes this leaf is bifurcated at the tip without the pinnae separating. Subsequent leaves have many pinnae. The adult leaf of the palm is unusual since the separate pinnae are not initiated as primordia, but arise as a result of ontogenetic tearing of the lamina. Two general patterns are evident depending on whether the terminal region of the leaf stalk carrying the lamina has completed its growth, or continues to grow as the leaf emerges from the bud. In the former case the segments of the leaf, the pinnae so-called, spring from a common origin, and the palm leaf has a fan shape, while in the latter the segments arise separately along the stalk, so giving to the leaf a feather-like appearance. These leaves should not be called palmate and pinnate.

The pinnae are sharply folded either with the fold up, *induplicate*, or the fold down, *reduplicate*. The pinnae meet the stalk in a characteristic manner, and along the stalk there may be prominent spines.

A crown of these leaves is formed at ground level and for any species the number of leaves in the crown is constant. There is continuous production of leaves, and as one senesces and dies another is produced. At first this young leaf is held vertically and the pinnae are tightly bundled together. At this stage it is called the *sword leaf*. The dead leaf may or may not be shed, but always the leaf base persists, and since this almost encircles the stem some protection is given to the axis by the persistent fibrous leaf bases. In many tropical areas this particular habit of the palm results in the production of a home for animals considered as vermin. The palm goes on producing leaves and losing them, there not being any periodicity of growth.

Except in the genus *Hyphaene* the stem is unbranched. If the growing point of a palm is damaged it may divide to give two separate growth centres, each producing an axis, and this can give the appearance of branching, but it is not a regular phenomenon in nature. Eventually the palm produces an inflorescence. This may be terminal; if so the palm is monocarpic, and after fruiting it dies. Some of these terminal inflorescences are massive, there being reports of some reaching 7 m in *Corypha*. Those palms which produce lateral inflorescences are polycarpic, and they may produce the inflorescences in bursts, though a large number form them regularly as they do their leaves.

The inflorescence that the palm produces is called a *spadix*, which implies that the axis carrying the flowers is fleshy, and is subtended by a bract which encloses the whole inflorescence in the early stages, the *spathe*. While the definition of a spadix includes fleshiness of the floral axis as one of its attributes, in palms the axis is more often woody, and certainly this is the most frequent state at maturity.

The individual flowers are small, regular, with an inconspicuous perianth of six members in two whorls of three. There are six stamens, again in two whorls of three, and three superior carpels. In the gynoecium various states are possible. All three carpels can be free and fertile, or the syncarpous condition with the three fused is found, or one of the carpels is fertile while the other two are sterile. In the syncarpous types only one ovule is produced, so at maturity the fruit is single-seeded.

The fruit may be a berry, nut or drupe. The pericarp may be prominent, and those with succulent fruits provide both man and animals with food. Usually the endocarp is hard and serves to protect the seed, but in some the endocarp is membranous.

In most palms the embryo within the seed is relatively small, and there is a massive endosperm, which in the case of coconut is liquid for part of the time. The seed is furnished with an embryo pocket, and covering this is an operculum which is less tough than the rest of the covering layers. It is through the operculum that the swelling embryo grows during the early stages of seedling development. A corresponding weak part is found in the mesocarp of those palms with hard fruit.

The palm, though woody, does not produce timber that can be used as lumber. On the other hand the palm does provide many economically important products. The young bud is prized as a vegetable, the so-called palm cabbage, but this is a very profligate use of the plant since on removal of the bud the plant is killed. Fibres are obtained from the leaves of some species, and from the leaves of others, the carnauba palms, a very hard wax is harvested. The pinnae can be used directly to weave mats and baskets.

Many palms, just prior to flowering, build up a considerable reserve of starch in the stem. They can be cut at this time and the stem processed to produce a starch of high quality, as in the sago palms. Nearly all palms have an abundant supply of sap to the inflorescence, and if this is removed the sap can exude from the peduncle stump. By 'tapping' the palm the flow of liquid

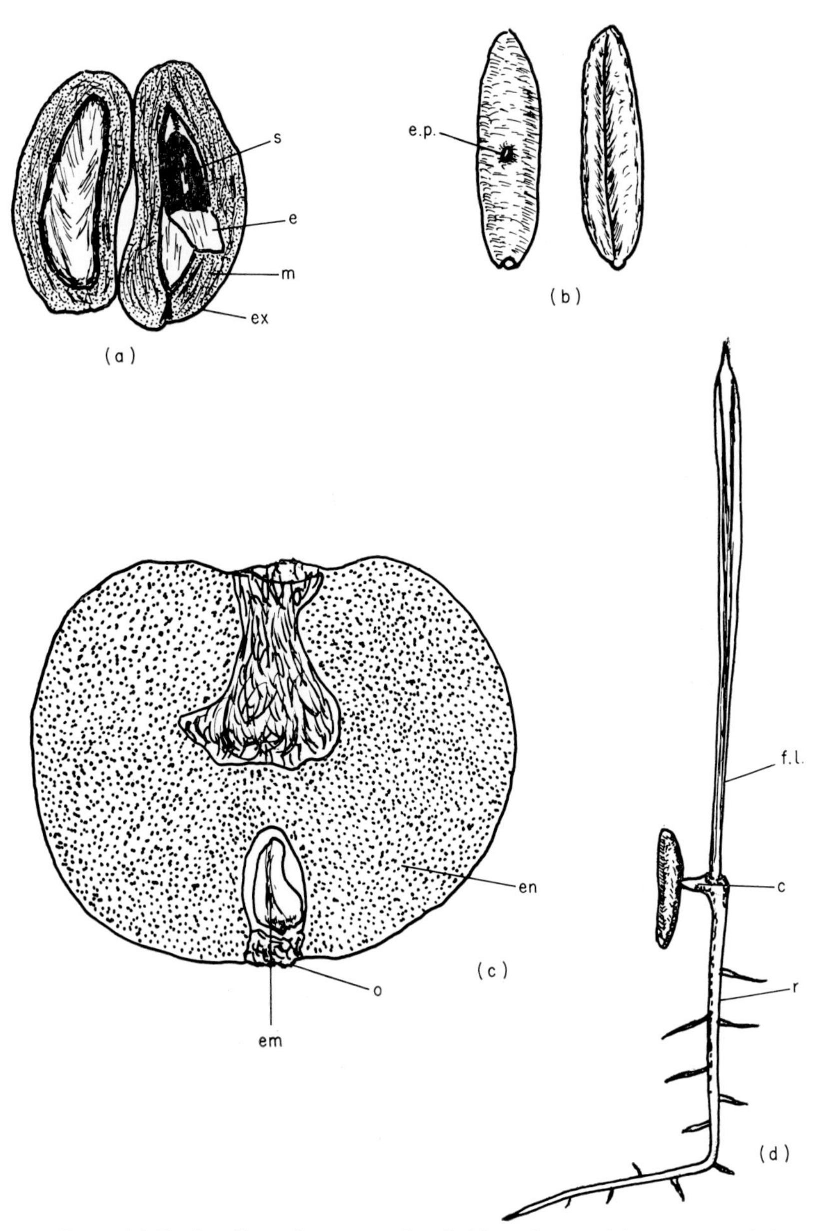

Fig. 29. Date. (a) Fruit split to show stone (seed) (s), endocarp (e), mesocarp (m), and exocarp (ex). (b) Seeds. (c) T.S. of seed: em, seed embryo; en, endosperm; o, operculum. (d) Seedling: c, cotyledon; r, radicle; f.l., first leaf.

can be maintained, and very high yields of a sugary fluid can be obtained. This liquid can be boiled down to a syrup, or even a crystalline sugar, but the most common outlet for this sap is the production of toddy, which is the fermented juice.

The fruits may be eaten directly as in date, and other fruits such as the betel nut are grown for chewing. However, the fruits of greatest importance are those which have high percentages of oil. The oil palm has a fruit in which both the pericarp and the endosperm are rich in oil, while coconut has in the fleshy part of the endosperm a material, *copra*, which is much prized for the oil it contains. Coconut fruits are valuable in another way because the mesocarp is composed of hard fibres and these are used in the production of coarse ropes and matting. The mesocarp fibres are called *coir*.

The endosperm of some South American species is extremely hard and a creamy white. The seeds are large and can be used as a substitute for ivory; indeed they are known as vegetable ivory.

The taxonomy of the palms is not based on natural divisions. Nine subfamilies are admitted, but in three of them there is only one genus and in one of these there is but a single species. Nowhere else in the plant kingdom is there such a situation of a single species with a subfamily recognition! Some authors have suggested that some of the subfamilies should be raised to the level of family, but the plants are so obviously palms that the suggestion has not been acted upon.

The widely grown and economically important palms belong to the subfamilies Phoenicoideae, Cocoideae and Arecoideae, but palms from the other subfamilies are locally important. The subfamilies are recognized by the type of folding of the pinnae, the nature of the leaf, feather or fan-shaped, the fruit structure, and whether the plants are monoecious or dioecious. Often in the former case, though the flowers are morphologically bisexual, they are functionally unisexual.

The important palms entering world commerce are *Cocos nucifera* L., the coconut, *Elaeis guineensis* Jacq. the oil palm, and *Phoenix dactilifera* L., the date palm. They are all typical palms with the habit and course of development described. Dwarf cultivars occur and are often preferred over the taller types mainly because of the ease of harvesting, this more than compensating for the lower yields attributable to these smaller plants. The date palm has been an item of trade for millenia, but the other two have only been important trading items within the last 150 years, though the coconut has provided tropical societies with many materials probably since the time man has lived in settled communities.

The Coconut Palm

This plant of the strand region is not known as a wild species. It is always associated with man and he has been responsible for its distribution throughout the tropics, particularly in the island complexes of the Pacific. It is a tall plant which may bear fruit within six years of germinating. The spadices are axillary and good plants will carry about ten inflorescences per year, there being a succession from those just opening to the most mature with ripe fruit.

The first-formed leaves are scale-like, simple and only after the third leaf do we observe the typical compound leaf of a palm. By the time the tenth leaf is forming, the characteristic feather leaf with reduplicate pinnae has been established and at maturity the leaves are carried in a tight crown of 25–35 leaves.

The stem is columnar, greenish, and there is evidence of the leaf scars all along its length. When growing near the sea the stem is curved outwards from the land and it is thought that this growth form is determined by a phototropic response of the stem to light reflected from the water. Plants on the landward side of a grove do not show this bending of the trunk unless there is an area of bare light-coloured soil next to the clump of trees.

The inflorescences are large, up to 2 m long, and when young they are covered by two sheaths, a smaller outer, and a larger inner, which is the true protective structure (the spathe). The smaller may fall away after being punctured by the larger, which in turn is torn from the tip down by the developing spadix. When the inflorescence emerges the spathe may fall off. The spadix is branched and carries numerous staminate and carpellate flowers, there being a preponderance of the former. The carpellate flowers are borne towards the base of the spadix. The male flowers are small and regularly trimerous, including a vestigial ovary with a trifid style. This rudimentary style carries nectaries. The larger female flowers are subtended by two bracteoles, have six perianth members in two whorls, but no evidence of stamens. In the centre there is the gynoecium of three fused carpels. Normally only one carpel produces an ovule but instances where each carpel develops an ovule are known. The single-ovuled gynoecium gives rise to a single-seeded fruit which is technically a fibrous drupe.

Coconuts (Fig. 30) have the following features. They consist of the endocarp and seed, the fibrous mesocarp and leathery exocarp having been discarded. The entire fruit can weigh as much as 2 kg, while large coconuts are about 1 kg. The pericarp is some 10 cm thick, the great bulk of it being the fibrous middle region, which makes the fruit buoyant. After harvest this mesocarp, along with the exocarp, is removed and constitutes coir, which is used for cordage and matting. The central structure left after removal of the coir is the seed, which has a very delicate testa, surrounded by the endocarp. At the base of this there are three large papillae. These represent the opercula for the emergence of the seedling. Only one of these has an embryo behind it and it is softer than the other two which are as coriaceaous as the rest of the endocarp. The relatively small embryo is associated with a large endosperm, which is made up of two types of tissue, colloquially referred to as water and meat. The meat is a layer 1 cm thick under the testa, and the water, or milk as it is referred to in Western countries, occupies the seed cavity. There can be as much as 400 ml of

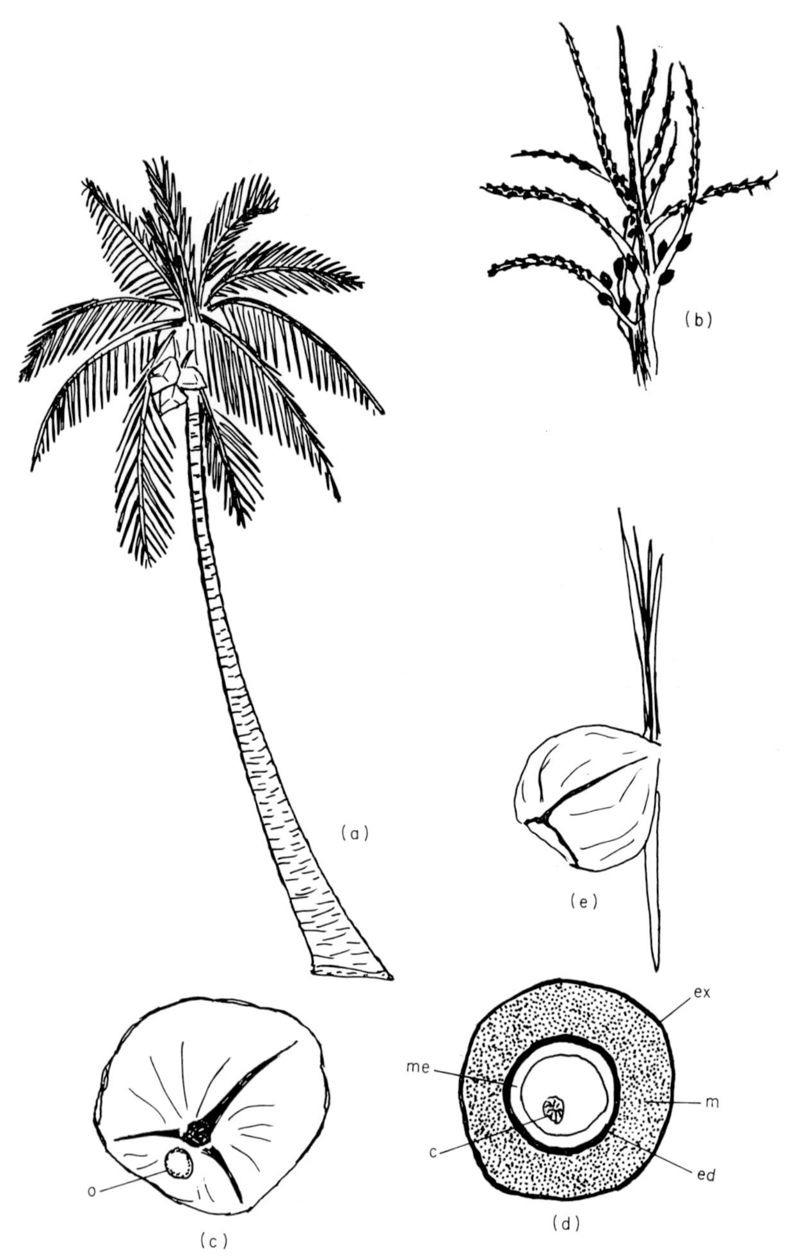

Fig. 30. Coconut. (a) Adult plant. (b) Spadix. (c) Fruit with operculum, o. (d) Section of fruit: ex, exocarp; m, mesocarp; ed, endocarp; me, meat. A coconut apple, the haustorial tip of the cotyledon, has formed (c). (e) Seedling.

this liquid. The liquid is eventually lost on storage, and the meat becomes gelatinous. The meat is the copra of commerce. Copra is prepared by splitting the nut and allowing the meat to dry and shrink away from the endocarp. After a while, perhaps two days, the meat is separated from the shell and dried as rapidly as is practicable to a moisture content of about 6%. This takes five to seven days. The dried copra can be stored and may have an oil content of 65%. Coconut oil, used in margarine production, is obtained from the copra by any of the traditional oil extraction methods and the residue is of potential value as a feeding stuff.

The Oil Palm

This is also a cocoid palm and therefore it has a form like that already described for coconut palm. The terminal crown is somewhat larger than the preceding, with 40–50 leaves. Like *Cocos*, *Elaeis* is monoecious, but here the usual circumstance is that an inflorescence is either staminate or carpellate. Occasionally individual flowers may be hermaphrodite, or an inflorescence will carry both types of flower, but the determination of sex of the flowers seems to be environmental though there is a genetic basis which is expressed by the relative number of female inflorescences which are produced. Commercial cultivars should have a satisfactory sex ratio with large numbers of female inflorescences.

The flowers are basically like those of *Cocos*.

The fruit is a sessile drupe, which can weigh up to 30 gm. The exocarp is hard, the mesocarp pulpy, and the endocarp horny. The mesocarp provides palm oil, and this in the early years of exploitation was the only oil marketed. Later it was found that the kernel, equivalent to the coconut, was also rich in oil, and this is extracted as palm kernel oil. At maturity the amount of fat in the mesocarp is from 70–75% of the dry matter and the composition of this oil makes it more valuable for the production of soap or for industrial uses than for margarine. The palm kernel oil is an edible oil. The residue left after extracting the kernel is a valuable cattle food.

The Date Palm

The date palm has been cultivated from ancient times for its edible fruits. It is dioecious and the first recorded instance of deliberate pollination relates to this plant. Because of the chemical composition of the fruit (up to 70% sugar at maturity) the plant can only be grown in regions with low atmospheric humidities, otherwise the fruits serve as a growing medium for saprophytic microorganisms. However, the plant will only grow where the water table is not less than two feet below the soil surface, hence we find the best situation for growth and production of dates around water in desert regions.

The Phoenicoideae are the only palms with feather leaves with induplicate pinnae. There is only one genus, viz. *Phoenix*. The date palm has a large crown with 100–120 leaves. The trunk may be as short as two metres but it can be as high as thirty.

The spadices are large, branched in case of male inflorescences which are erect and broom-like. The individual staminate flowers are cream-coloured with six perianth members and six stamens. The female flowers have the six perianth members typical of the family and the fruit is large drupaceous. The pericarp is fleshy. The seed is cylindrical pointed up to 2.5 cm long and deeply grooved. The small embryo is in a median embryo pocket opposite the groove. Though germination is easy this palm is vegetatively propagated using suckers which form at the base of the adult palms.

CHAPTER 5

The Musaceae: The Bananas

Amongst the fruit crops of the world the banana is of prime importance. Purseglove states that only in the case of grape is there a greater production but it is likely that a greater weight of bananas is moved in trade than of any other single fruit. In the tropics this plant is certainly the one which yields the greatest amount of edible starch per hectare, and is deservedly the greatest of the fruits of the hot regions of the world.

The family Musaceae as described by older systematists includes a large number of giant herbs, but more recent treatments of this group of monocotyledons have tended to confine the family to the genera *Musa* and *Ensente*. These two genera are distinguished from the others that were formerly included within the family by having their leaves arranged spirally, whereas the other genera had opposite leaves, and in having the posterior perianth member free with the others fused.

The modern treatment of the genus *Musa* has been definitively stated by Simmonds, and while this is different from that adopted in most other cultivated genera, the peculiar circumstances surrounding the origin and selection of bananas justifies what might be considered a unique taxonomic approach. In their native habitats bananas are found on disturbed soils such as along river banks and on scree slopes. They have, as we shall see, a good method of vegetative reproduction, and an ability to form polyploid hybrids. This particular combination of circumstances provides the opportunity for the development of clones which are morphologically distinct but genetically related, but in the absence of seed production it is difficult for the genetic affinities to be demonstrated. Most of the cultivated forms are seedless.

Bananas are large herbs up to 6 m high with the leaf sheaths giving a hollow pseudostem. This morphology is similar to that of the grass, and, as in that plant, the flowering shoot elongates rapidly growing within the hollow leaf sheaths until the infloresence is exerted. One significant difference is that the stem of the banana is naked. Since the production of seeds in the commonly grown types is rare, the plant is propagated vegetatively. The unit of propagation is a sucker, the '*sword shoot*' which is planted out to grow on to give the plant which will be ready for harvest some nine to 18 months later.

When planted the sucker has enveloping its bud an outer layer of scale leaves. On establishment, leaves which are upright and rolled in the bud develop, and though the lamina expands, the leaf does not become noticeably pendant. These leaves are sometimes known as *sword leaves*. Eventually the leaves that are produced have a pronounced sheathing base, encircling the axis at the point of insertion, which narrows before terminating in a massive blade. The region from which the blade is subtended is referred to as the stalk. The blade is up to 4 m long by 1 m broad and has a pronounced midrib. The veins arise at right angles to the midrib and some 2 cm apart. This large lamina is rolled in the bud and after expansion it droops characteristically. Even gentle winds will tear the lamina between the veins, and soon the blade becomes quite tattered. Any one leaf remains functional for a set time and then the lamina along with the stalk is lost though the base remains. At any time after establishment there is an approximately constant number of leaves per plant, for as one leaf is lost another is produced to take its place (cf. palms).

As with all plants there are buds in the axils of the leaves, but only those at the basal nodes develop. They grow out as short sympodial rhizomes, sending up a number of shoots around the mother plant. The subterranean part of the sucker axis is swollen and may be likened to a corm, for it is this which affords the propagule. The suckers which surround the mother plant are of different types, and while any can be used in the establishment of a banana plantation it is considered that those showing sword leaves, the 'sword shoots', are the best. Some suckers just erupt through the soil and are covered only in scale leaves. These are the '*peepers*' and others, '*maidens*' have expanded adult type leaves, and may even produce inflorescences along with the mother plant.

At flowering a large terminal compound inflorescence is produced. In *Musa* the flowers are carried in clusters, each cluster being subtended by a large brightly-coloured spathe. The groups of flowers are carried on the main axis in the form of a raceme so that the flowering head is quite a massive structure (Fig. 31). While the individual flowers may each have the essential parts represented there is a sharp division into the basal ones which function as females and the terminal clusters which are essentially of male flowers.

The structure of the flower in this family shows a considerable degree of complexity, and in *Musa* itself the flowers are considered representative of an advanced form of monocotyledon. Each flower is asymmetrical and the zygomorphy is associated particularly with the perianth, though in the male flowers there is modification of one of the stamens so establishing zygomorphy in the androecium. The perianth consists of two whorls each of three members, but the outer whorl along with the two anterior inner perianth segments are united into a tube at the base with a strap-shaped terminal portion. The inner posterior perianth member is free.

In the male flowers there are five functional stamens, the sixth, the inner

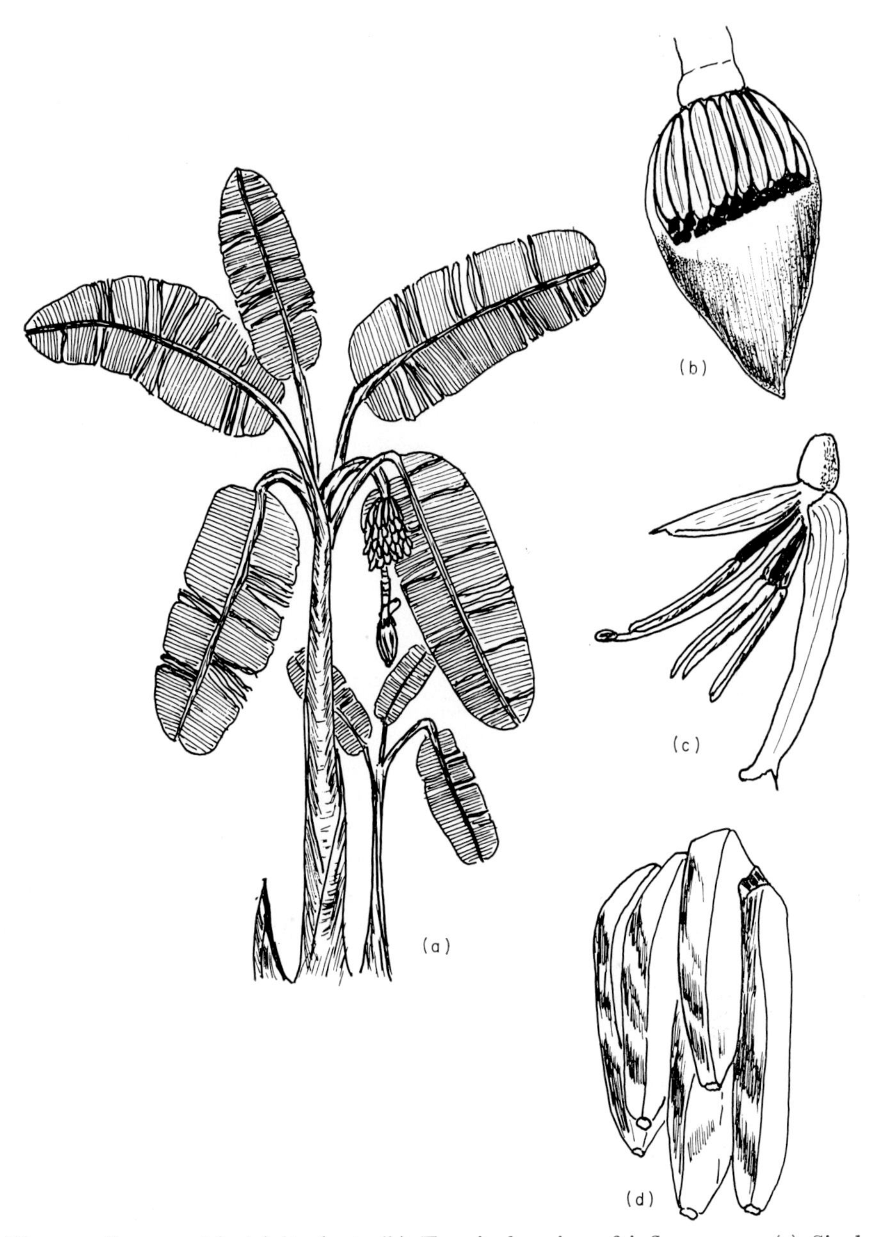

Fig. 31. Banana. (a) Adult plant. (b) Terminal region of inflorescence. (c) Single male flower. (d) Ripe fruit.

posterior, being represented by a staminode. Finally in these flowers there is a small abortive ovary complete with a vestigial style and stigma. This ovary is inferior.

Female flowers differ from the male in having five staminodes and a functional ovary of three carpels; again, this structure is inferior. There is a robust style and a large trilobed stigma.

The cultivated bananas produce fruit parthenocarpically, but the wild bananas require pollination for the production of fruits. Occasionally seeds are formed in the cultivated species but this is a rare event.

The taxonomy of the cultivated bananas is confused because of the production of clones. Simmonds proposes that in the bananas there should be adopted a procedure for specifying the genomic constitution of the morphologically distinct clones and that the simple Latin binomial should be abandoned. There is some justification for this since the species which have been described do not represent natural taxonomic units as is evident when hybridization studies are undertaken.

The genus is divided into five sections and the cultivated bananas belong to the section *Eumusa*. The cultivated types are found to have the chromosome numbers 22, 33 or 44 and are therefore either diploids, triploids or tetraploids. This polyploid series is considered to be based on two wild species *M. acuminata* Colla, and *M. balbisiana* Colla which have been given the genome designations AA and BB respectively. Straight *M. acuminata* diploids are encountered as cultivated forms but the AAA triploid is more vigorous and has larger fruits. The diploid hybrid AB is hardier than those clones which have been derived exclusively from *acuminata* parents, and can be grown in regions with seasonal climates. Triploid hybrids, either AAB or ABB, have arisen and the incorporation of the B genome makes them hardy, which along with their greater vigour and large fruit confers upon them the desirable features of a crop plant. Simmonds proposed nomenclature for the bananas would therefore be e.g. *Musa* (AAB group) 'Horn Plantain' referring to the genus, the genomic constitution and the cultivar type respectively. This is a sound pragmatic way of naming the bananas and this practice could with advantage be applied to other polyploid series where the plants reproduce vegetatively.

The much smaller *M. textilis* Nee belonging to the section *Australimusa* was once grown widely for its leaf fibres. These fibres were marketed as Manilla hemp which could be manufactured into ropes and cordage. Like all of the minor fibre crops which were grown for specialized purposes the world production is falling because synthetic fibres can be tailor-made for the specific purposes which could at one time only be met by crops like Manilla hemp.

CHAPTER 6

The Cucurbitaceae: Melons, Cucumbers and Gourds

This is a family of the warmer regions of the world. Representatives are found in both the Old and New Worlds with seven genera common to both. There may be as many as 100 genera and 850 species, all of which are intolerant of frost. One genus, *Byronia*, does reach to north-west Europe.

The members of this family are annual, or rarely perennial, climbing or straggling herbs. They grow rapidly and the tissues are sappy, with abundant air spaces. The internodes and petioles are hollow. The plants climb by tendrils which arise at the nodes adjacent to the insertion of the leaf. The tendril is seen not to be truly axillary. The exact nature of this organ has been the subject of much controversy, but since there are many buds in the axil it is possible that the tendril is derived from a displaced bud, or even an extra-axillary bud, and is in fact a modified axis. It has at various times been thought of as homologous to root, leaf, stipule or flower stalk. It has even been considered to be a unique structure not related to any other plant organ.

The leaves are spirally arranged, and are in most cases markedly palmately veined. The blades may be entire, lobed, or almost dissected. In the genus *Citrullus* the leaves are pinnatifid.

The flowers which are diclinous are carried in the axils and may be solitary but are frequently in inflorescences of various types (Fig. 32). Various patterns of distribution of flowers are found. Most of the cultivated species are monoecious, but in cucumber sex expression can be varied both genetically and environmentally so that plants which are wholly female exist and the species might be thought to be gynodioecious. Staminate flowers are usually produced first in the basal axils and these are typically pentamerous, and in both the calyx and corolla there is fusion of the parts giving a well-defined tube. The petals are often large and showy, with yellow or yellow-orange the most commonly encountered colour. The stamens can be considered as having arisen from five primordia but there is much irregular fusion during development, the anthers not only fusing but also becoming contorted and proliferated. This type of flower is stated to have syngenesious stamens. In the male flowers

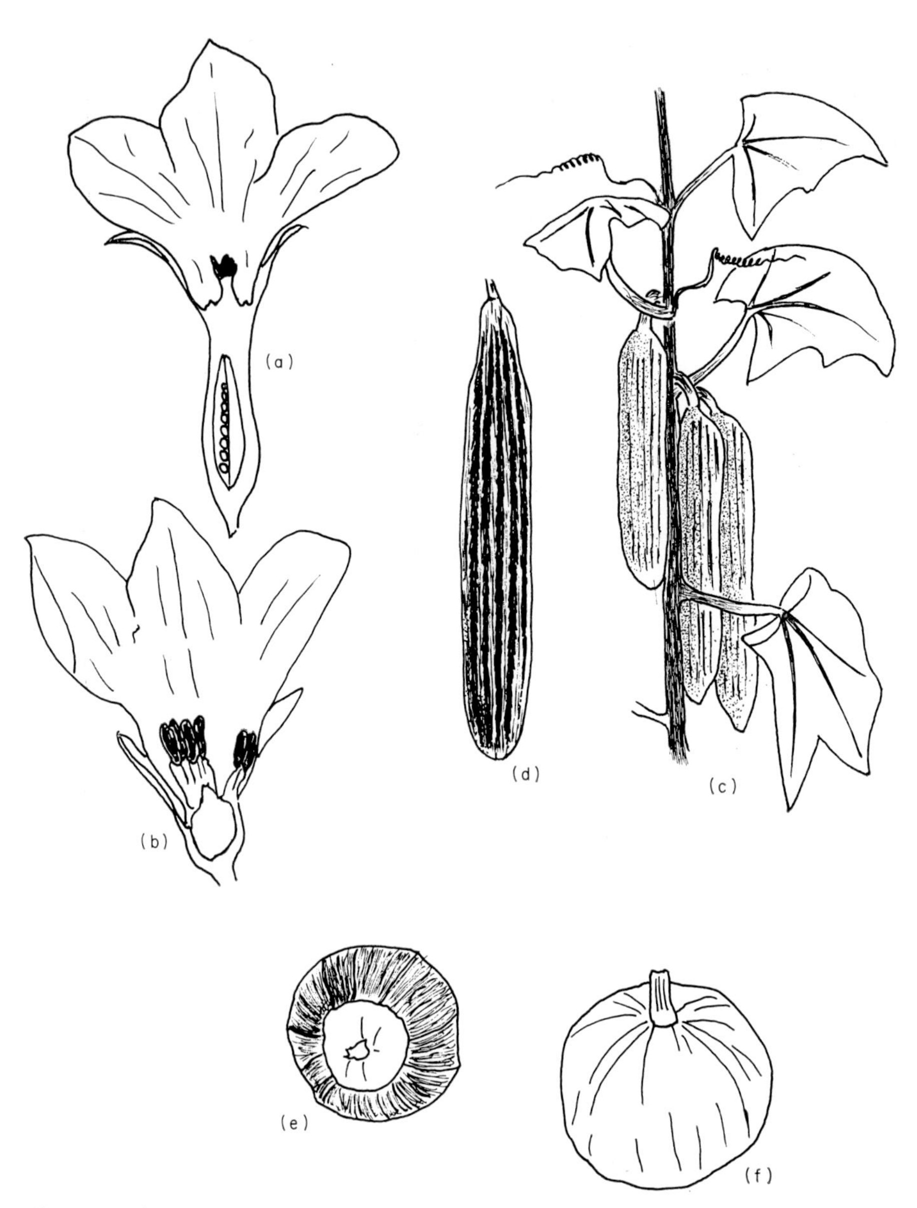

Fig. 32. (a) Female flower of cucumber. (b) Male flower of cucumber. (c) Portion of cucumber plants with developing and mature fruits. (d) Cucumber fruit. (e) Fruit of *Cucurbita moschata*. (f) Fruit of *Cucurbita pepo*.

there is a rudimentary style. The outer parts of the female flower, the calyx and corolla are like those of the male but stamens are absent though staminodes may be present. The multicarpellary ovary is inferior and is surmounted by a short robust style which ends in a stigma with as many lobes as contributing carpels. The cultivated species are tricarpellate. Pollination is by bees.

The fruit is fleshy but with a hard rind that is derived from the receptacle. This inferior berry-like fruit is termed a *pepo*. It can be very succulent and in desert regions the fruit can substitute for water, for example the edible portion of the water melon contains up to 95% water.

The most important genera are *Citrullus*, *Cucurbita* and *Cucumis*, in which the fruits are major items of trade, and of lesser importance because of being grown for local consumption are *Luffa*, *Lagenaria* and *Sechium*.

Citrullus lanatus (Thunb.) Mansf., the water melon, is cultivated throughout the tropics and subtropics. *Citrullus* differs from the other cultivated members of the Cucurbitaceae by having pinnatifid leaves, and this feature alone serves to identify it (Fig. 33). The plant is a coarsely hairy annual, which may either be grown as a climber or allowed to straggle. If grown as a straggler the climate must afford a dry season at the time the fruits are ripening, otherwise they will rot on the ground. This particular problem is present in all these climbing herbs and in intensive agricultural systems it is better to grow the plants on wires but in the case of low cost crops, outdoor cultivation, in which the plants are allowed to grow over the ground, is preferred.

Water melon is a variable plant, the leaves showing varying degrees of lobing, but it is in the size and shape of the fruit that the greatest variation is found. Small spherical fruits are favoured in some regions, while in others large ellipsoidal fruits are the types which are marketed. Some of the small fruits have a diameter of 25 cm while the largest fruits might be 60 cm long and weigh up to 20 kg.

In the genus *Cucumis* two species are widely grown, *C. melo* L. and *C. sativus* L., the melon and cucumber respectively. These species are similar but the melon has short petioles, about 5 cm long while the petioles of cucumber can be 15 cm. Cucumber has stiffer hairs and the flowers tend to be smaller. Perhaps the fruits are the most distinctive feature separating the species. Melons have fruits which are not substantially greater in length than in breadth whilst cucumbers are much longer than they are broad. There is considerable variation of fruit forms and these have given rise to the accepted market types of both melon and cucumber. In Europe and the United States the musk-melon with smooth to delicate netted rind, and the Cantaloupe with a heavily sculptured rind, dominate the market, but in Asia, the Casaba, with a smooth rind parti-coloured green and yellow, along with types with more elongated fruits, are popular. The cultivar 'Honey-dew' is a Casaba type melon and it is imported into Europe as the 'winter' melon.

Cucumber has long been grown under protected conditions in Europe and the cultivars used in this practice have long smooth fruits with tender skins.

These types have come to be called English or forcing cucumbers. They contrast with the field cucumbers which have smaller fatter fruits covered with spines, and possessing a thick bitter rind. Small fruited varieties are grown for pickling fruits and can be classified as gherkins.

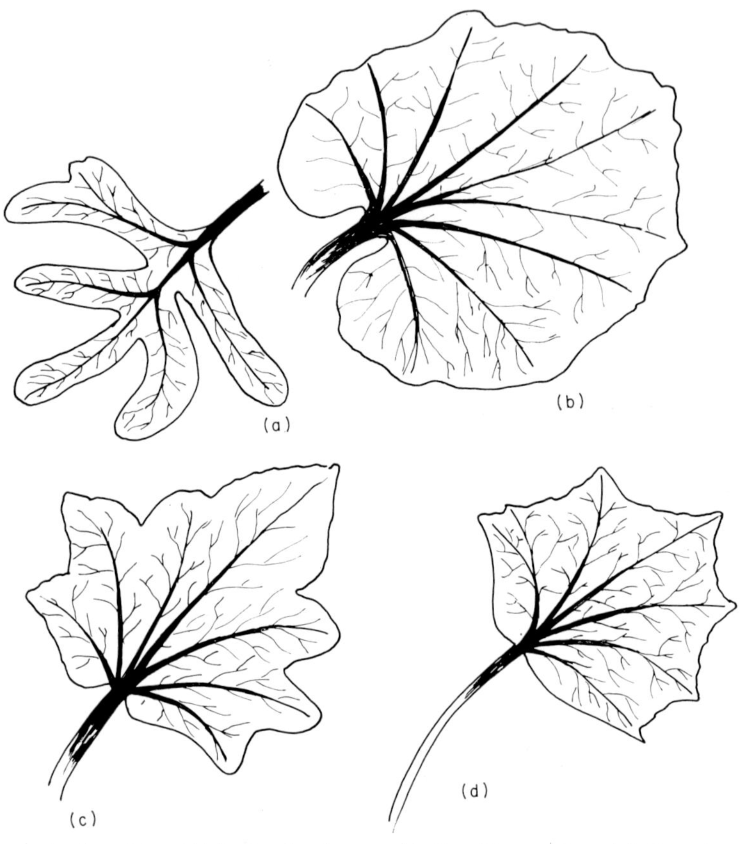

Fig. 33. Leaf profiles of (a) *Citrullus lanatus* (b) *Cucurbita moschata* (c) *Cucumis melo* and (d) *Cucumis sativus*.

The forcing of cucumbers has been the subject of considerable work by both the plant breeder and the plant physiologist. This is not surprising when it is realized that the gross return on an acre of glasshouse-grown cucumber can be considerable. Most of the intensively grown fruit is produced parthenocarpically, and this in itself overcomes all the problems associated with pollination in a monoecious species. A gene for all-female flowers exists in *C. sativus* and this has been incorporated into the most popular cultivars. Seed of these all-female cucumbers is expensive, but the certainty of production of fruits more than compensates for this high cost. The possession of this gene is particularly valuable to those growers producing fruit under long-day conditions, since photoperiods in excess of 14 h promote maleness. The promotion of maleness is interesting, because the plants which are tending towards the production of male flowers also exhibit longer internodes, pronounced tentril formation, smaller leaf blades, and are usually lighter green. This collection of features can be induced by an application of gibberellic acid. The opposite syndrome, towards femaleness, short internodes, large leaf blades, fewer tendrils and a darker green colour can be induced by application of dwarfing compounds. Whether physiological manipulation of the habit of the plant will be widely undertaken when there are many cultivars which are all-female, is a matter which is under consideration.

The genus *Cucurbita* has provided us with the squashes, pumpkins and marrows. The taxonomy at the species level has been confused because of the range of variation met. It is possible to recognize four species—*C. pepo* L., *C. maxima* Duch. ex Lam., *C. moschata* (Duch. ex Lam.) Duch. ex Poir., and *C. mixta* Pang. *C. pepo* has hard stems and leaves with robust spiculate bristles whilst the others have either soft stems with few bristles, *C. maxima*, or without bristles. *C. moschata* can be separated from *C. mixta* by its peduncle which has a large flaring button at the point of attachment to the fruit.

The colloquial name squash is used indiscriminately for the fruits of these plants. Large fruited forms of *C. pepo* are termed marrows, whilst the large-fruited forms of the other three species are the pumpkins or winter squashes. The small-fruit types, harvested green for immediate consumption, are given names appropriate to the shape of the fruit, and small-fruited marrows are called courgettes or zucchini.

CHAPTER 7

The Cruciferae: Cabbages and Turnips

Many of the crop plants have had an ancestry from families which have centres of distribution within the Mediterranean basin, or the Far East or Central South America. The Cruciferae exemplifies this as well as any other. Another feature found with some of the crops is that the progenitor was a plant of the sea coast, and again we find that the most important cruciferous crops are derived from wild species which are naturally distributed on sea cliffs and on sandy soils inland.

The family is one found in the cooler higher latitudes and has few representatives amongst its 300 or so genera and 3000 species anywhere in the tropics or southern hemisphere. Most members are herbaceous, though small shrubby forms are present. Annuals, biennials and herbaceous perennials are all well represented, and in the case of the monocarpic species there is usually formed a rosette from which the flowering axis is produced. Other species are caulescent from the beginning. The plants are often hairy and all, including the glabrous species, have a characteristic smell and bitter taste. The smell and flavour arise from cell constituents and not from any glandular hairs which might be present. When all the features of the family are considered its members are found to be very closely allied, and this family is one of the more natural that is recognized. Indeed some of the genera are so close that intergeneric hybrids can be produced with ease, and while this indicates genetic affinity between taxa, the ease with which genes can be exchanged can result in some difficulties with classification.

The seeds of the Cruciferae are rich in oil which may be used as an edible or fixed oil, though at one time the seed oils were used mainly as illuminants. As well as containing these oils, which are chemically triglycerides, the members of this family have in their seeds, and other parts of the plant to a lesser degree, mustard oils. Chemically these are isothiocyanates, i.e. R—NCS (R = an organic radical) and the compound may be present as a glucosinolate which on hydrolysis by the appropriate enzyme releases the free mustard oils. All isothiocyanates are pungent, irritant substances, which are present to excess in the wild species, such that the plants are decidedly unpalatable. By selection, the level of mustard oil has been reduced and the vegetative parts of the crop species are now quite palatable, though a diet too rich in cruciferous plants may lead to flesh and milk taints. Man, on the other hand, has deliberately selected some species for the pungent oils to produce a condiment. The mustard oils are not considered fatally poisonous though instances of poisoning by horse-radish have been reported.

On germination the seed produces two obcordate cotyledons, which come above ground and expand. The apical notch of these organs is pronounced, and except for a few species, the cotyledon shape is usually sufficient to permit an identification of the seedling to the level of the family. The first leaves follow the expansion of the cotyledon, and while the older leaves may be entire, the younger leaves have a much-dissected lamina. There are species and sub-species where all leaves are entire. The development of a divided lamina in this family is due to the midrib growing more rapidly than the wing with the result that portions of the lamina are left all along the stalk. The terminal part is large and is often an inverted lyre-shaped blade. That the portions of the lamina along the stalk are not equivalent to pinnae is clear from their asymmetrical distribution and shape. The last-formed leaves on the flowering stem are reduced and may have clasping bases.

The inflorescence is a much-branched raceme and the flowers are ebracteate (Fig. 34). The individual flowers are most frequently actinomorphic, but in some genera where the inflorescence is a corymb the outermost flowers of the head have zygomorphic corollas due to enlargement of two of the petals. The flower is basically dimerous, the parts being arranged in twos. In this respect the family is like the others of the order Rhoeodales.

The outer pair of sepals are median, and with the inner laterals make up the calyx. These inner sepals have saccate bases into which is secreted nectar. The

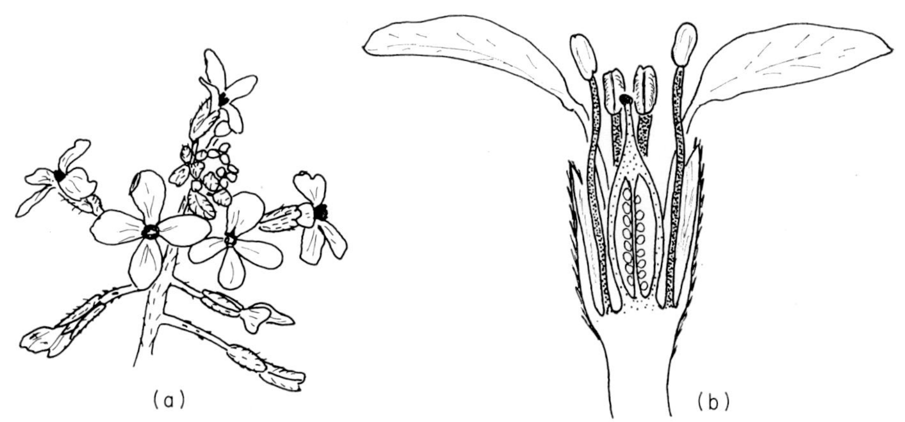

Fig. 34. Radish. (a) Inflorescence. (b) Half-flower.

corolla of four petals is in a single whorl but the petals alternate with the members of the two whorls of the calyx. The stamens number six, a lateral pair with short filaments, and four median being composed of two sets each of two, each set alternating with the petals. These inner stamens have long filaments and the flower is said to be *tetradynamous* on this account. These inner stamens arise not from four primordia but from two which after inception split, so giving rise to what may be thought of as twinned stamens. The pollen is shed inwards as a result of the anthers splitting along an inner line. The gynoecium is of two fused carpels arising from a well-developed disc which secretes nectar. The bases of the filaments are also associated with this disc. There are two carpels fused to give an ovary with parietal placentation. During the course of development there arises from the placentae an ingrowth which eventually results in the formation of a septum so giving two loculae at maturity. This is the *false septum* or *replum*. The fruit (Fig. 35) is a capsule which dehisces from the base up leaving the seed attached to the replum. Two forms of capsule are produced. In the first the fruit is as broad as it is long, and this is called a *silicula*, while in the second the fruit is considerably longer than broad, and the terminal portion is free of seeds. In this case the fruit is termed a *siliqua*, and the terminal part is called the *beak*. In some species the fruit is a schizocarpic *lomentum*.

There are no subfamilies in the Cruciferae, and the tribes are established on the presence or absence of glandular hairs, whether the stigmatic surfaces are more prominent over the placental regions, and the way in which the cotyledons are folded in the seed, which at maturity is without endosperm. The only tribe with economically important genera is the Brassiceae. Amongst the others many genera have been selected as ornamentals.

Brassica is the most important genus and from this one genus there have been selected the most important of the green vegetables of the temperate zone.

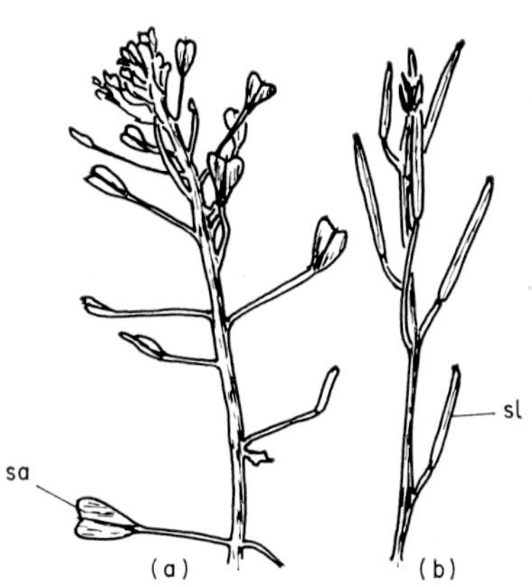

Fig. 35. (a) Fruiting head of shepherds purse (*Capsella bursa-pastoris* L.) showing siliculas. (b) Fruiting head of hairy bitter cress (*Cardamine hirsuta* L.) with siliquas.

These same species have also been subject to selection to give fodder crops with very high levels of digestibility, up to 85% of the plant being digested by the ruminant, and low residues, with the result that an animal can ingest large amounts and give good live weight gains against the amount of plant dry matter consumed.

Two wild species, still found in the native floras of north-west Europe and the Mediterranean, can be improved over as few as ten generations to give a plant that is commercially acceptable. These are the wild cabbage, *Brassica oleracea* L. and the wild turnip, *B. campestris* L. Recognizably distinct crop plants derived from these two wild species have at times been given full specific rank, but it is now accepted that such crop types should be referred to the above species and their distinctive morphological features should be used to establish taxa at either the level of variety, or perhaps better, forma.

In their general growth habit the species, *B. oleracea* and *B. campestris*, are very similar but there are significant differences which are carried through to the crops which have been selected from them. Both species may be either annual or biennial in habit, though with *oleracea* the tendency is towards the biennial while with *campestris* annuality seems to be the norm. In the early juvenile phase the plants develop as rosettes and only in the reproductive phase does a distinct stem develop. In the wild cabbage the rosette may be preceded by there being elongation of a few of the internodes. The wild cabbage has shiny glabrous blue-green leaves, while the wild turnip has grass-green leaves which are covered in pronounced hairs, especially the older leaves. The wild species has the characteristic leaf shape described earlier. Such a leaf might be said to exhibit ontogenetic tearing of the lamina. In some of the cultivated forms this feature is not seen and the leaves are entire.

During the time the rosette is forming the rootstock is enlarging to become a storage organ and in the wild turnip the upper portion of this rootstock can be substantial. In the wild cabbage the rootstock is noticeably woody. The hypocotyl contributes to this storage structure and, depending on the relative amounts of hypocotyl and root involved in the formation of the storage organ, so is determined how deeply the structure is buried in the soil.

The storage organ, colloquially called a 'bulb' (Fig. 36), is produced by an abnormal secondary thickening. The root is typically diarch and very soon a cambium is laid down between the primary phloem and primary xylem. Cambial activity can begin by the time the first true leaf is expanding. The main tissue formed as a result of the early cambial activity is a mass of fibres with vessels as secondary xylem, but only enough to result in the cambium adopting a circular profile in transverse section. Whenever this has happened the secondary xylem that is then formed consists of short radial rows of vessels with large medullary and vascular rays of parenchyma. The tissues of the rays give rise to islands of meristematic cells which proliferate, so enlarging the rays and disrupting the regular arrangement of the radially oriented vessels. This results in the vessels appearing in their ordered rows only near the cambium. The

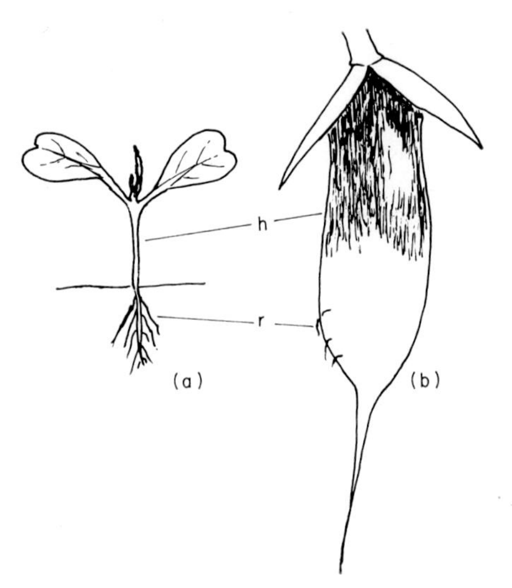

Fig. 36. Radish. Seedling and adult storage organ showing how the latter is derived from the hypocotyl (h) and the radicle (r) of the former.

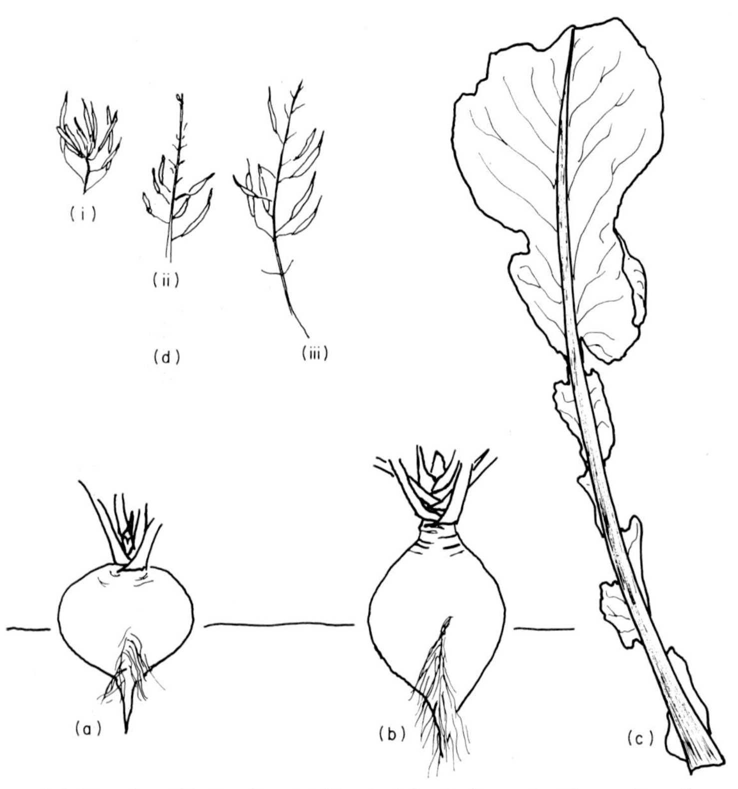

Fig. 37. (a) Turnip. (b) Swede. (c) Typical leaf of turnip illustrating the ruptured lamina. (d) Terminal portions of infructescences of (i) turnip, (ii) swede and (iii) cabbage.

secondary phloem is not as abundant as the secondary xylem, and there is not the same disruption of the zones of sieve tubes by proliferating ray tissue.

The cortex is split at points opposite the protoxylem and is sloughed off. A periderm is produced from a cork cambium that is laid down in the pericycle. This cork cambium gives rise to a smooth outer protective tissue which is not well supplied with lenticels.

From the wild turnip there has been selected two main types of plant. In the first the 'bulb' has enlarged to provide us with the commercial turnip, but in the second the plant has been selected to give a much-branched stem. These latter are the rapes which may be grown either for their leaves or their seeds. The wild cabbage, on the other hand, has not provided a form with an enlarged root-stock, all the cultivated forms being selections in which the stem and/or buds have been enlarged or modified.

Along with the species of *Brassica* just considered there occurs the wild mustard *B. nigra* Koch. This is an annual which does not produce a storage root-stock. Nonetheless, this species has played an important part in the evolution of the crop forms derived from the genus *Brassica*, and in its own right is of some commercial importance in the production of mustard.

The taxonomy of the cultivated *Brassicas* has been confusing, but as a result of hybridization studies the following nomenclature has been adopted:

Brassica campestris
Without 'bulb': forma *annua*, annual rape (turnip-like)
forma *oleifera*, winter or oil-seed rape (turnip-like)
With 'bulb': forma *rapifera*, turnip

In this context the term turnip-like means that the plants have hairy grass-green leaves. As well as these forms widely grown in Europe and America, there are some grown in the Far East as leaf vegetables. Older terminologies included turnip as *B. rapa* L. and the Asiatic leaf vegetables as *B. chinensis* L. and *B. pekinensis* Rupr., but the morphological variants should be considered as forms rather than as subspecies or species.

Brassica oleracea
With blistered leaves: forma *bullata*
(i) terminal bud enlarged as perched rosette: subforma *subauda*, Savoy
(ii) axillary buds enlarged: subforma *gemmifera*, Brussels sprout
(iii) large plant with open terminal bud: subforma *palmifolia*, palm cabbage

Leaves not blistered
(i) terminal bud enlarged, perched rosette: forma *capitata*, cabbage
(ii) large open plant much branched: forma *acephala*, kale
(iii) large plant compact with terminal branches leafy: forma *ramosa*, kale
(iv) stem swollen: forma *gongyloides*, kohl-rabi

(v) peduncles, pedicels, or flower buds enlarged: forma *botrytis*
inflorescence open: subforma *cymosa*, sprouting broccoli
inflorescence compact, leaves widespread: subforma *cauliflora*, cauliflower
inflorescence compact, leaves overarching: subforma *italica*, broccoli

The swede is a plant similar in many respects to the turnip. It is hardier and has a higher dry matter content, and in some ways seems intermediate between cabbage and turnip (Fig. 37). It differs from turnip in having blue-green leaves which have hairs mainly over the veins. The stem does elongate, so that above the swollen rootstock there is a definite neck. This plant is *Brassica napobrassica* D.C. and we now know that it is the amphidiploid between *B. campestris* and *B. oleracea*. This hybridization must have occurred soon after man took the genus into cultivation, though it is quite possible that the parental species had produced a hybrid which had established itself as a wild plant. Indeed, there is a set of species in which hybridizations have taken place and we know that the hybrids agree with taxa that have been established and accepted as good species in their own right. The relationship between these species and their hybrids is shown in the scheme outlined below. The appropriate somatic chromosome numbers are given in brackets after the species.

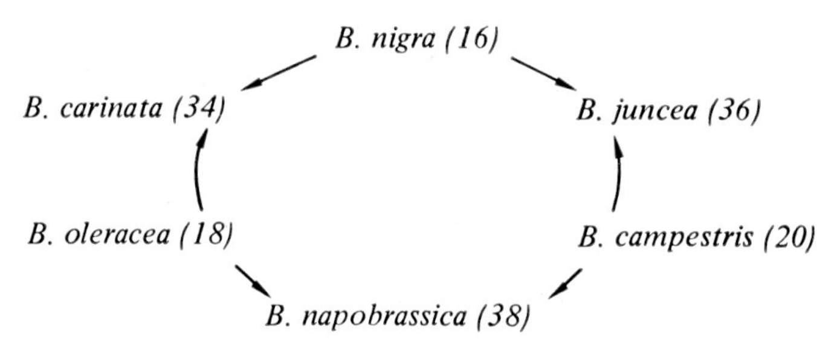

The arrows point from the parents to the hybrids.

Both *B. juncea* (L.) Czern et Coss. and *B. carinata*, A. Brown tend towards the annual *B. nigra* rather than the other parent and in fact *B. juncea* is as widely grown for condiment mustard as is the original black mustard.

Our description of these crop brassicas has concentrated on the features which have been selected for their economic value. Other differences exist, particularly in the inflorescence.

While the flowers are typical of those of the Cruciferae, the genus is recognized by the upright sepals which do not reflex at any time prior to their dropping off. This distinguishes *Brassica* from *Raphanus* where the sepals are markedly reflexed. In cabbage the petals are lemon-yellow and the flower is relatively large, almost 2 cm across. Turnip has a small (1.5 cm or less) flower which is a deep orange-yellow. The hybrid nature of swede is seen in its flowers

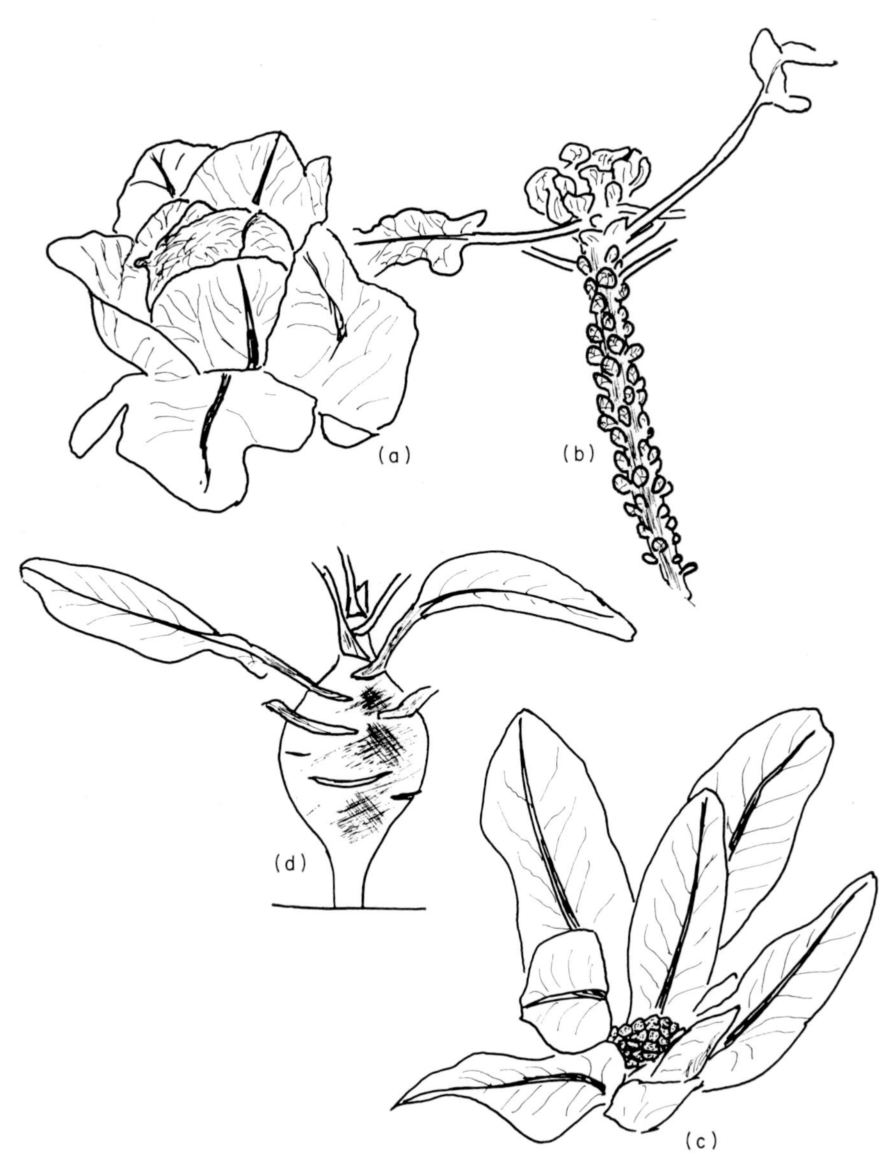

Fig. 38. Forms of *Brassica oleracea*. (a) Cabbage. (b) Brussels sprout. (c) Cauliflower. (d) Kohl-rabi.

being intermediate. When in fruit there is produced a siliqua with a beak, the sizes corresponding to the flower sizes. However, in cabbage the stalks (peduncles) of the lowermost fruits do not elongate, so the infructescence maintains a pyramidal shape. In turnip these lowermost fruit stalks elongate to carry the bottom fruits to a level above that of the upper ones, so that the infructescence is overtopped. Once again the swede is intermediate. These inflorescence characters are most useful in identifying wild plants and some of the leaf vegetable forms.

Before leaving the brassicas the detailed nomenclature of the swede should be considered. Some have called the yellow-fleshed types of swede rutabagas, keeping the term swede for those cultivars with creamy or white flesh. Swede exhibits the same types of morphological variation as its turnip ancestor so we have

Brassica napobrassica
Without 'bulb': forma *annua*, annual rape (swede-like)
forma *oleifera*, winter or oil seed rape (swede-like)
With 'bulb': forma *rapifera*, swede

The term swede-like here refers to the leaves. *B. napus* L. refers to these swede-like rapes and in fact these are the ones that are now most widely grown. Oil seed rape is a crop of increasing importance since it is the best oil crop for climates with low summer temperatures. Rape or colza oil was at one time produced solely as an illuminating oil. This was because these older types of rape produced an oil which had high levels of erucic acid in them. The newer cultivars have been selected for low levels of this compound and rape oil is now considered an edible oil. Unfortunately the residue cannot be used as cattle food without the possibility of its causing digestive upsets.

Among the minor vegetables belonging to the Cruciferae there is radish, *Raphanus sativus* L. which has a form like that of a miniature turnip, but has a peppery-flavoured rootstock; the flower has spreading sepals, and the fruit is inflated, with a few seeds and a large beak. Horse radish, *Armoracia lapathifolia* Gilib. is another which is a perennial producing a fleshy tapering root. This root can penetrate to 60 cm and when harvested and grated gives the condiment horse-radish sauce. The uncooked residue from this plant is poisonous to pigs, and due care should be taken if hotel kitchen refuse is fed to these animals.

There are many weeds and ornamentals belonging to this family, but none possesses any unusual characters except members of the genus *Iberis* which have corymbose inflorescences with the outer flowers zygomorphic. This exception emphasizes the natural grouping of the family.

CHAPTER 8

The Chenopodiaceae: The Beets

Beets have been grown as vegetables for many years but it is only within the 20th century that a selection, the sugar beet, has become one of the most important crops of the cool temperate regions. The production of sugar is dominated by two crops—sugar cane in the tropics and sugar beet elsewhere. Because of transport costs and the relative ease of mechanizing the production of sugar beet, beet sugar is competitively priced in comparison with cane sugar, though in many countries governmental policy allows for subsidization of the crop.

The Chenopodiaceae is a comparatively small and distinct family. Most of the members are found associated with a particular ecological situation, viz. an elevated level of salt in the soil. Plants which can grow satisfactorily in such saline habitats are called *halophytes*, and while it should be recognized that the most important adaptations for a plant to live successfully in such a habitat are physiological, halophytes may possess characteristic morphological features. In the Chenopodiaceae these morphological adaptations are seen as succulence and the possession, in many of the species, of mealy glands covering the surfaces of the leaf. The halophytism of beet is seen in the way in which the plant responds to the application of common salt as a fertilizer. Whether the plant really requires the sodium ion or whether the effect the salt has on the crop yield is due to the depressing of the weeds normally found in beet crops, is a matter which has not been satisfactorily resolved, but the addition of common salt is said to result in higher sugar yields, 100 kg salt per hectare giving an additional 100 kg sugar per hectare, with this correlation holding to an application of 750 kg salt per hectare, after which the salt has a detrimental effect. It should be remarked that there are many soils which cannot tolerate an application of sodium since it results in the formation of sodium clays which are most intractable.

One of the physiological features of this family of halophytes is the plants' ability to withstand high levels of inorganic ions within the cell sap. Among the ions that are found in much higher quantities than normal are nitrate and nitrite. Both of these, but particularly nitrite, have the property of causing scouring in animals, and nitrite is also thought to be involved in the formation of certain nitrosoamines which have been stated to be carcinogenic. The type of digestive upset caused by these ions is of little consequence in an adult but there did occur in the United States an outbreak of diarrhoea in infants which was eventually traced to batches of spinach which had been grown under very high levels of fertilizer nitrogen.

The deposition of calcium oxalate is in many plants considered to be a mechanism of detoxification. Beet and other chenopods often contain this compound, and since it is toxic to mammals, care must be exercised when the green parts of beet are used as vegetables or for fodder.

Most members of this family are herbaceous, but a few are shrubby. Among the more unusual types are the glassworts, in which there is reduction of the leaves and the stem is modified as a succulent water-storing organ.

The natural distribution is confined to certain coastal regions, and internal saline deserts. There are perhaps no more than 75 genera with some 700 species.

The Mediterranean Basin is one of the recognized centres of distribution, and two plants from this region have been selected as crops: the biennial *Beta vulgaris* L. and the annual *Spinacia oleraceae* L. from which has been derived spinach. From the American centre, *Chenopodium ambrosoides* L. and *C. quinoa* Willh. have given us the Mexican tea and quinoa respectively. From the first is extracted an excellent vermifuge, but in quinoa the seeds are used for the production of a flour.

The wild beet is found on the sea cliffs of north-west Europe and while Linnaeus considered it to be distinct from the common beet, it is now considered that the two plants belong to the same species. If any distinction is to be made, the establishing of subspecies of *B. vulgaris* is permitted. The cultivated beet then becomes *B. vulgaris* subsp. *vulgaris* while the sea beet is *B. vulgaris* subsp. *maritima* (L.) Thell.

The beet (Fig. 39) is a biennial, though in some races the annual habit predominates. The seed germinates slowly to give a seedling with strap-shaped glossy cotyledons. A rosette of large, simple glossy leaves with robust midribs and a succulent petiole is produced. In some instances the cultivars are heavily pigmented with the red pigment betacyanin. This water-soluble plant pigment is restricted to a few families, all in the order Centrospermae, and it should not be confused with the widely occurring red anthocyanins. In fact, the possession of betacyanin is a good taxonomic character.

While the rosette is developing there is produced a robust rootstock which in the cultivated forms becomes fleshy. In the wild sea beet this rootstock is woody. At flowering the plant becomes caulescent producing a much-ribbed stem which branches profusely. The primary inflorescence consists of a branching raceme but the flowers are carried in small cymes; there may be up to six flowers in each cyme.

In beet the flowers are simple monochlamydous, the perianth having five members. The five stamens are opposite the perianth and there is an ovary, with two or three divisions to the style, partly embedded in the receptacle. The ovary contains a single curved ovule attached to its base. The actual number of carpels in the ovary is probably three. After fertilization the perianth enlarges and becomes corky. During the period of enlargement of the perianth, adjacent fruits coalesce so that at maturity a cluster of structures, based on the secondary inflorescences, is produced. Each unit of the cluster is strictly a false

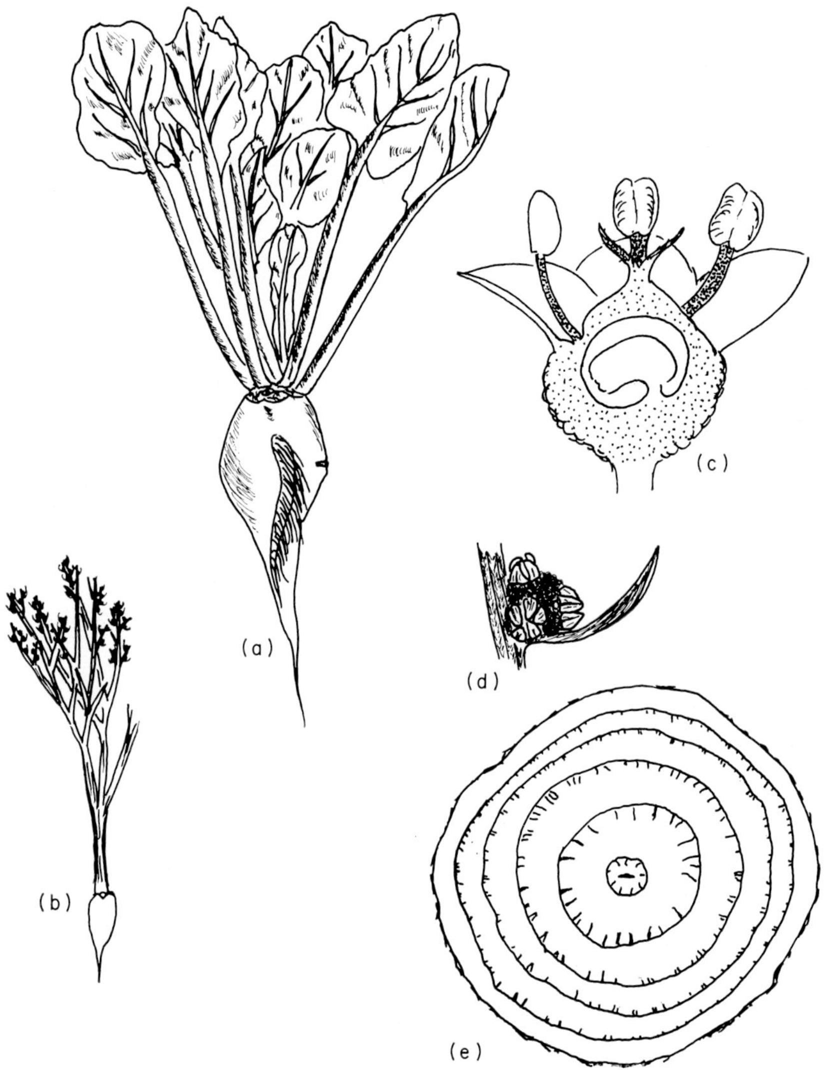

Fig. 39. Beet. (a) Adult plant. (b) Steckling. (c) Half-flower (d) Fruit cluster. (e) T.S. of root.

fruit, so the cluster can hardly be called an infructescence, although it is functionally similar to one.

The beet cluster presents a problem in the production of the crop, since the seeds within the cluster usually germinate over a period and the seedlings are held together within the corky mass of the cluster. Both these features prevent easy singling to give one seedling per station, which is highly desirable. Modern precision drilling requires that the seeds or fruits should be uniform, and this is not the case in beet. Scarification of the clusters can be carried out to release the true seeds but this is a costly and imprecise procedure. The other alternative possible is to breed new cultivars which are genetically homozygous for the character of one flower per inflorescence. At maturity such clusters are single-fruited, and are termed *monogerm*. This second method, of obtaining a single seed at each station, is preferred.

The wild beet has been the ancestor of four main root crops—garden beet, sugar beet, fodder beet and mangold, and has also given rise to two leaf vegetables—spinach beet and swiss chard.

Of the root crops, the garden beet tends towards being annual, whilst the other three have a biennial habit though amongst any of the beets there are individuals which flower in the same year as sown. These 'bolters' are troublesome since their roots become woody and a high incidence of them in the sugar beet crop can reduce the yield of sucrose considerably. The production of bolters is determined by both the genotype and the climate.

Beet is a plant which requires both vernalization and exposure to long days in order to flower. Some beets are responsive to low temperatures at a very early stage of development, with no more than three adult leaves formed, and if such seedlings are exposed to prolonged cool dull days in early spring they will become ready to respond to the long days of summer and will flower in the autumn. The incidence of bolters is higher in the more northerly areas of the crop's distribution.

Any plants which produce flowering shoots in the autumn and overwinter are called *stecklings*. These stecklings harbour aphids which serve as vectors for many viruses including that of beet yellows, so the production of bolters should be prevented.

The root stock is produced by what might be an anomalous type of secondary thickening encountered only in a few families. Like the turnip the first signs of secondary thickening are seen in the very small seedling, almost as soon as the first true leaf has expanded. The root is diarch, and the cambium is laid down as an arc between the primary xylem and phloem. The two arcs expand to join opposite the protoxylem. Cambial activity is initially most intense at the pinches of the arc and the first products are usually xylem fibres. Some secondary phloem is also formed. This first cambium is followed by others which are originated within the pericycle and phloem, both primary and secondary. As each cambium is formed it produces a small amount of secondary xylem and some secondary phloem, with the subsequent cambia forming in the outer regions of the last-formed secondary phloem, which always has mixed within it derivatives from the pericycle. As many as 30 of these cambia can be produced. A well-defined periderm is eventually established.

During the early period of secondary growth the cortex is sloughed and the very small stele can be exposed. If climatic conditions are adverse this small stele can be subject to desiccation with the possibility of death of the seedling. This condition is known as strangles.

The hypocotyl also participates in the formation of the storage organ, but in the different crops, and types within a crop, there are different degrees of the contribution made by the hypocotyl to the final root. In the garden beet of the globe type, nearly all of the root is derived from the hypocotyl, whilst in the sugar beet this structure hardly contributes to the storage organ proper. Mangold has a series from globe, through tankard, to long, in which the hypocotyl contributes progressively less. Ease or difficulty of harvesting is related to the amount of root that is involved in the final storage structure produced. The more root there is included, the more difficult the beet is to remove from the soil. It has not been possible to breed globe types with high sugar contents.

In spinach beet the lamina is much enlarged and leaf production is prolific. Swiss chard, on the other hand, has been selected for large succulent petioles and midribs. Both of these forms are locally important.

The true spinach, *Spinacia*, is an annual with smaller, roundly hastate leaves which are not as glossy as those of beet.

There are many pernicious weeds in this family which can act as host for the causal agents of the diseases found in sugar beet. Amongst the worst are *Chenopodium album* L., fat hen and *Atriplex patula* L., common orache or iron root.

CHAPTER 9

The Malvaceae: Cotton

The Malvaceae is a moderate-sized family with perhaps 50 genera and 1000 species distributed worldwide. All habits are found, there being herbs, shrubs and trees, and since the flowers are somewhat showy many species have been taken into cultivation as ornamental plants. However, the economic importance of this family lies with those species which are cultivated for their fibres, and cotton is the most important fibre-producing plant in the world. The genus *Hibiscus* has a species which is grown for fibre and has also, in *H. esculentus* L. provided one of the important tropical vegetables, okra.

The Malvaceae is a very natural family and its members are easily recognized by the form of the flowers. Generally the plant has stipulate leaves alternately arranged, and with prominent palmate venation. Various degrees of lobing are found, and in some leaves there is almost complete dissection between the veins resulting in a leaf which is almost compound. Hairs are prominent, and these are most abundant on the stems. The flowers may be solitary or in compound cymes, each individual flower being subtended by pronounced bracteoles. These bracteoles have been considered by some authors to be a modified epicalyx, but it would appear that they are truly bracteate and not part of the flower proper as would be the case with an epicalyx. It is often thought that the epicalyx represents the stipular part of the sepal primordium and reflects the supposed homology of that flower part with leaves. Any family in which stipules are prominent might be expected to have an epicalyx in the flower. The bracteoles vary in number and it is interesting to note that where the bracteoles are few the odd sepal is anterior, but where the bracteoles are many the odd sepal is posterior. The flower is pentamerous, and we find that the sepals may be free or fused. The five petals are convoluted in the bud so that when the flower opens they are seen to be overlapping in a characteristic way. The stamens are unusual; the outer row is missing but the inner whorl is represented by a staminal column which is tubular at the base. The petals are associated with this column and at first glance it appears as though it might be the petals which are forming the tube. From the upper portion of the staminal column a large number of filaments is produced and each of these is surmounted by a pollen-carrying structure. These structures are unlike the typical anther and most can be thought of as a single pollen sac or theca. Usually the androecium is considered as having an indefinite number of stamens but the number of primordia that contribute to the formation of the column should be taken into account, and not the number of thecal heads. The gynoecium varies within the family from a single carpel to an indefinite number, fused to give a single ovary. The ovary has as many loculae as contributing carpels, except in one of the tribes where the ovary is divided transversely into single ovuled compartments. The single style is surmounted by a multifid stigma, the number of divisions again reflecting the number of carpels in the syncarpous ovary. Most flowers are protandrous, but in the absence of pollination the stigmas grow in a recurved fashion to bring the receptive surfaces in contact with the pollen shed from the stamens of the same flower, so ensuring pollination.

The ovary is multiovuled, and the seed usually has the embryo curved and lying in an endosperm. Some of the seeds are rich in oil. The fruit is normally a capsule, but in okra the capsule is harvested while the pericarp is still fleshy. A schizocarp is found in some species.

The family is divided into four tribes, but only one, the Hibisceae, is of consequence in having members that produce economically important crops.

Cottons have been grown for generations to provide man with a soft resilient fibre which can absorb moisture without losing its character, unlike most of the artificial fibres which are unable to absorb moisture, or wool, which can, but the wet garment loses its shape. The fibre is the hair produced in the surface of the seed coat. All cottons belong to the genus *Gossypium*. This genus is distributed throughout the tropics and representatives are found in all the continents. There are some 30 spp. but only four are grown commercially.

Gossypiums are subshrubs to small trees in habit and while many are perennial, the most popular cultivated species are annuals or are treated as such. Small dark glands are abundant over the surface of the stems, less so on the leaves, but are very prominent on the cotyledons. As well as these glands the plants are covered in hairs, to a greater or less extent. The branching pattern in the cottons is unusual. At each node there are two buds, but only one grows out to give the lateral branch. In the lower part of the plant these laterals are monopodial and purely vegetative, but in the upper parts of the bush the buds that grow out exhibit a sympodial habit and carry flowers. It has been claimed that the buds are distinctive, and that there is a commitment to a particular growth form before the bud has expanded.

Each flower is subtended by three large bracteoles, each with a much-toothed tip. These teeth may extend quite deeply down into the bracteole. These bracteoles are persistent and remain at fruit maturity. The calyx is cup-shaped and the five large petals with spreading terminal regions overlap in their basal regions to provide a corolla which has what is effectively a tubular base. There is a disc-shaped nectary at the junction of the corolla and the staminal tube and this nectary, along with three extra floral nectaries at the top of the peduncle near the bracteole, serves to attract bees, or even humming birds in the case of the American species. The androecium is typical for the family, and in the gynoecium there are usually three carpels, sometimes more, giving a three-celled ovary. As the seed develops, hairs arise from single cells of the seed coat. The hairs may be long with cellulose deposited spirally on the wall, and at maturity the cell lumen is still present; or there is produced a short hair with much cellulose thickening on the wall, the lumen being insignificant when the hair is mature. The long hairs are called *lint*, while the small ones make up the *fuzz*. Lint is easily removed from the seed but the fuzz adheres strongly, but even if removed it is found that the fuzz cannot be spun. Lint hairs can be up to 4 cm long, and considering that the hair is unicellular this represents a considerable degree of extension for a single cell. The wild *Gossypiums* are lintless and some of the cultivated types do not possess fuzz. The seeds are rich in oil and after removal of the lint this oil can be expressed to be used as an edible oil, with the residue providing a good cattle cake.

The four cultivated species can be divided into the Old and New World cottons. The Old World cottons have been assigned to *Gossypium arboreum* L. and *G. herbaceum* L. while the New World species are *G. barbadense* L. and *G. hirsutum* L.

The origin of the cultivated cottons should be considered within the framework of the cytological groups which are found in the genus. The wild lintless species can be divided into six sections within which there are four distinct genomes. These cytological groups are geographically distinct and crossing within a group is easy but hybridization involving different genomes is difficult. All the wild cottons except *G. tomentosum* Nutt. ex Seem are diploid, *G. tomentosum* being tetraploid.

The genomes are designated A, B, C, D and E. The genome A has been allocated to the Old World cultivated cottons, B to those species which have been found in Africa, C to those found in Australasia, D to the cottons of Central and South America, and the genome E to the species native to the deserts stretching from the Sind to the Sudan.

It is probable that the Old World cultivated diploids have been selected from a wild diploid progenitor in which the mutation for lint bearing had arisen. The most probable region for this to have happened seems to have been in Africa, and as a result of migration, perhaps assisted by man, the linted character spread towards Arabia and India. Cultivation and geographical isolation from the wild parent would enable the development of enough genetic isolation to result in the cultivated species eventually becoming possessed of a distinct genome.

The New World cultivated cottons are tetraploids with the genomic constitution AADD. How the A genome was transferred from the Old World to the New is not known. It seems as though only cottons with the A genome are those which can give rise to linted species.

Within each of the cultivated species there is a number of recognizably distinct races. They may be distinguished by their morphology and their persistence. The species are most easily recognized as follows, but since crossing is widespread, seeds collected from plants growing along with other species, as is common in botanic gardens, may not breed true.

Bracteoles with shallow teeth, often blunt
  Bracteoles spreading with 6–8 teeth, capsule rounded: *G. herbaceum*
  Bracteoles erect, teeth 3–5 or entire, capsule pointed: *G. arboreum*

Bracteoles with long, lacinate teeth
  Leaves deeply lobed, lobes 3–5, capsule densely glandular: *G. barbadense*
  Leaves lobed, to half depth of blade, usually 3 lobes, capsule smooth: *G. hirsutum*

The cultivated races of *Gossypium herbaceum* are: *africanum* (Watt) Hutch. and Ghose, *acerifolium* (Guill. and Perr.) Chev., *persicum* Hutch., *kuljianum* Hutch., *wightianum* (Tod.) Hutch. Those of *G. arboreum* are: *indicum* Silow, *burmannicum* Silow, *cernuum* (Hutch. and Ghose) Silow, *sinense* Silow, *bengalense* Silow and *soudanense* (Watt) Silow. The New World tetraploids are likewise found to have distinct races, and in *G. barbadense* these are *brasilense* (Macf.) Hutch. and *darwinii* (Watt) Hutch., while in *G. hirsutum* the races recognized are *marie-galante* (Watt) Hutch., *punctatum* (Schum.) Hutch. and *latifolium* Hutch.

In areas of high-cost labour cotton is now grown as an annual, rather than as a short-lived perennial. The annual forms are considered as more advanced, but the main reason for preferring the annual habit in this crop is the ease with which it can be harvested. Usually the crop is defoliated by chemical treatment to allow harvesting by stripping or spindle picking, as a once-over operation by machine. Where hand labour is available and the cost is competitive, hand picking at intervals is carried out, and the yield is higher. Some three or four pickings are taken in the harvest, the first picking beginning six months after sowing. In picking, the ripe capsules, called *bolls*, are removed as they dehisce, so exposing the seeds covered in the hairs.

The actual time of flowering, and therefore fruiting, is determined by genotype and the environment. The plant is a short-day type but the expression of flowers is not possible until the plant has reached a stage when the sympodial buds will grow out to produce inflorescences. Even though these buds are present at all of the nodes, some varieties have the propensity to produce sympodial branches low on the main stem, while others only produce them in the upper part of the shrub. However, the production of fruit is as much under the control of an abscission mechanism in which flower buds are shed unopened. This shedding is most pronounced if the soil is too wet. Selections in which both the photoperiodic sensitivity and the abscission mechanism have been subordinated have been made, and these provide not only the annual cultivars,

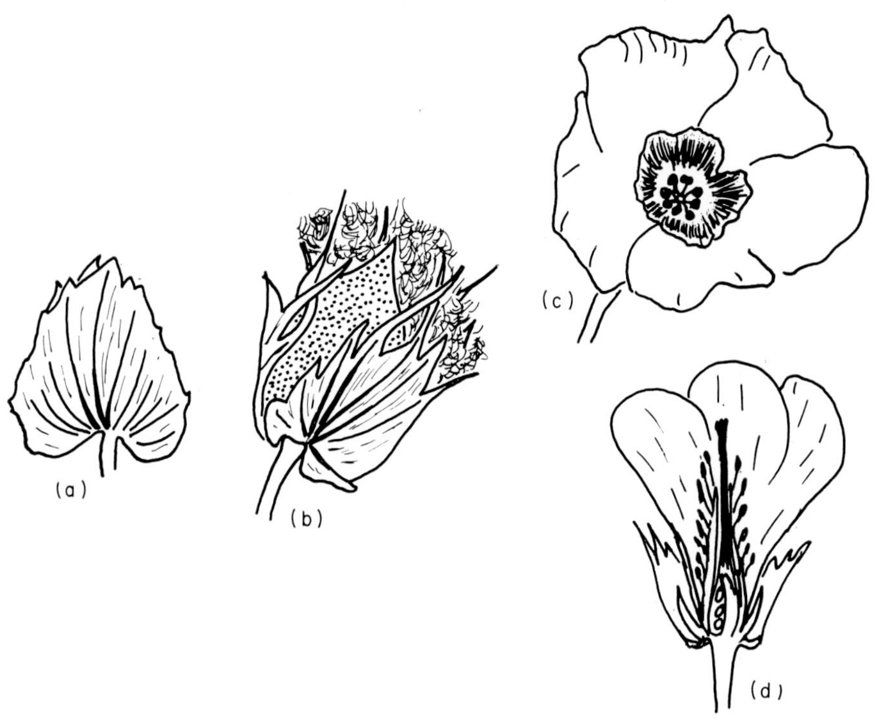

Fig. 40. (a) Bracteoles of *Gossypium herbaceum*. (b) Opening boll of *G. barbadense* with persistent deeply-toothed bracteoles. (c) Flower. (d) Half-flower of *G. barbadense*.

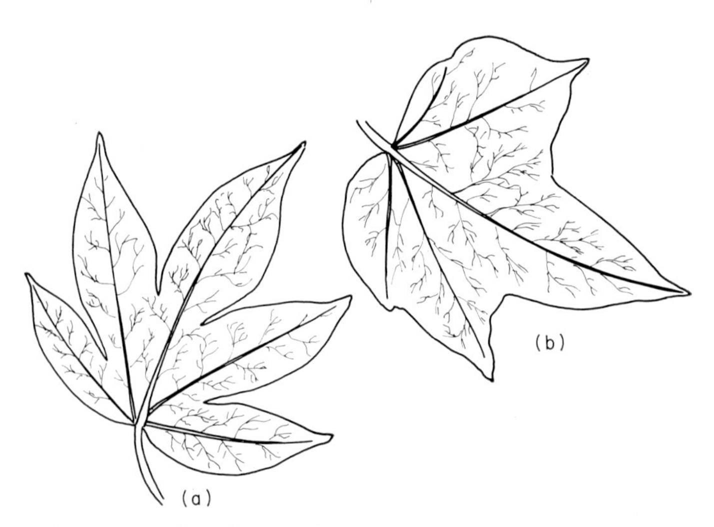

Fig. 41. Leaf profiles of (a) *G. barbadense* and (b) *G. herbaceum*.

but have also resulted in the range of the crop being extended to more temperate regions.

The genus *Hibiscus* has many similarities to *Gossypium*. Many produce stem fibres which are used locally but only in the case of kenaf, *H. cannabinus* L. do we find a crop which enters world trade. The product is Deccan jute. It is obtained from the stems in a similar manner to jute and the fibre has much the same properties, though coarser.

Hibiscus esculentus L. (*Abelmoschus esculentus* (L.) Moench.) okra or Lady's fingers, is a popular vegetable in some parts of the United States. It is a tall erect annual with palmately lobed leaves which have small stipules. Flowers are carried singly in the upper axils, and the fruit is held upright. This fruit is a tapering capsule with a beak portion free of seeds. It is harvested whilst still green and brittle.

CHAPTER 10

Miscellaneous Fruits: Citrus and Pineapple

The number of types offruit entering world trade in substantial amounts is small. Most fruits are grown locally for consumption by the indigenous population, but in highly urbanized and industrialized countries there is a movement of fruit to the centres of population. Since most of the industrialized countries are in the temperate regions we find that the fruits that are marketed have to travel and store satisfactorily. Only a few species produce succulent fruits with these characteristics.

The temperate fruits nearly all belong to the Rosaceae (q.v.), but in a few regions members of the Grossulariaceae and the Ericaceae are grown for their berries. Tropical fruits are numerous but here again we find only a few species which are grown for export. Occasionally 'exotic' fruits will be marketed, but usually have to be transported by air and this is reflected in the high prices charged for such oddities.

Bananas, oranges and pineapples are the most commonly encountered tropical and subtropical fruits. Because bananas are a staple food in the world context and not really a fruit crop to be thought of as a luxury, they have been considered elsewhere. Oranges and related fruits are never considered as staples, nor is the pineapple, and though these are profitable crops, the acreage grown could be reduced in the event of world food crises and the area released could be used to grow some staple crop.

Citrus Fruits

The orange is a member of the genus *Citrus*. *Citrus* belongs to the family Rutaceae which contains some 150 genera and 1600 species. Most of the species in this family are shrubs or trees, though herbs are encountered, and many exhibit xeromorphic characters. The leaves are usually compound, exstipulate, and arranged either alternately or opposite. Many possess glandular dots and nearly all are strongly aromatic.

The flowers are mostly regular and tetra- or pentamerous but other arrangements occur. The calyx may or may not be united below, with the odd sepal posterior. Alternating with the sepals there are as many petals, but the stamens, generally twice the number, exhibit *obdiplostemony*, i.e., the outermost row are opposite, not alternating with, the petals and these stamens are generally inserted on a disc which lies below the ovary. The ovary is superior and consists of five fused carpels. There is a single style with a terminal globose stigma. Fruits are either berries or capsules (Fig. 42).

As well as *Citrus* a few are grown for their essential oils secreted by the glands on the leaf, and others for their fruits.

The Rutaceae is divided into seven subfamilies and as far as we are concerned the subfamily Citroideac is the only one of importance. This subfamily has a single tribe, the Aurantieae which has 30 genera. *Citrus* is a member of a subtribe of six genera all very closely related. Intergeneric hybrids are obtained with ease as also is the capacity to form graft unions which means that *Citrus* scions

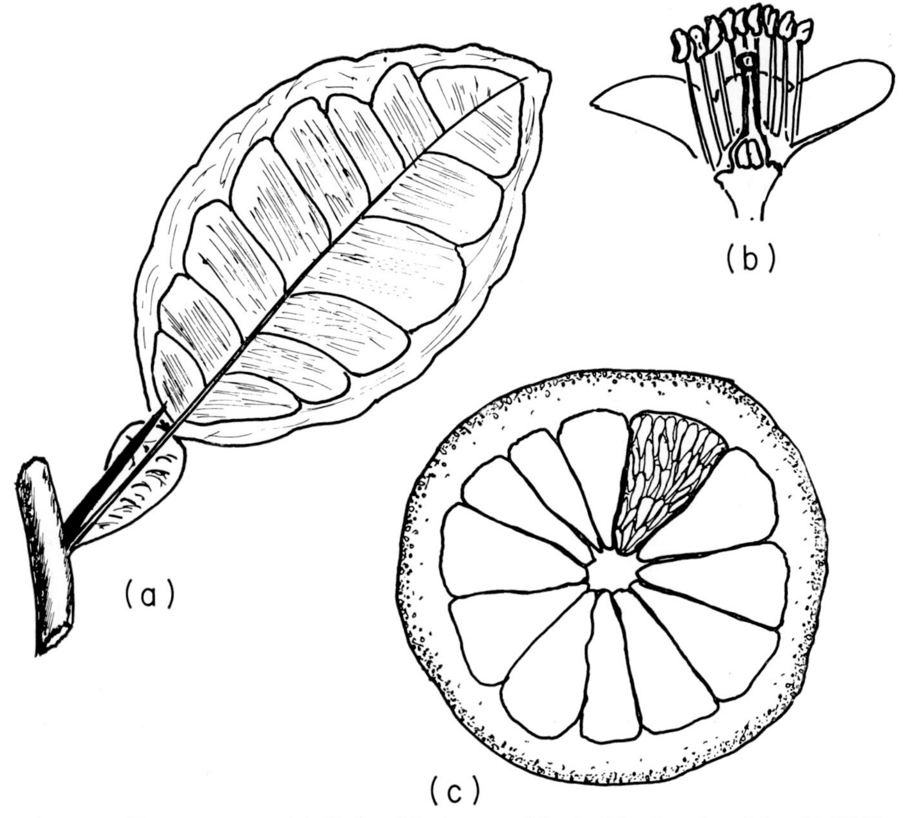

Fig. 42. *Citrus aurantium.* (a) Node with thorn and leaf with winged petiole. (b) Half-flower. (c) Section of fruit.

can be grown on any one of a number of rootstocks. Because of the widespread occurrence of interspecific hybrids the taxonomy of the genus *Citrus* is confused.

The genus *Citrus* has a natural distribution in the tropics and subtropics of India, South-East Asia and China but is now cultivated extensively in the warmer Mediterranean climates of the world. The most important regions for commercial production are Spain, North Africa, Israel, South Africa, Florida, California and Brazil. All species are small evergreen trees with leathery, glossy leaves in which the petiole is winged and with a distinct joint between the petiole and the blade. There are no stipules. At each node there is a thorn, seen most prominently on wild species, having been selected against in cultivation. The thorn is considered to be the modified first leaf of an axillary bud which has grown out.

The inflorescences are axillary corymbs carrying fragrantly scented flowers. Much variety is shown in the flower and it is different from that described for the family. Sepals may be four to eight in number as also the petals, and in the androecium we find the stamens in irregular bundles but arranged in a single whorl. Generally the stamens are indefinite. Similarly there is an irregular number of carpels and occasionally a second whorl of carpels is produced to be incorporated in the final syncarpous ovary. When this occurs the fruit is like that of the 'navel' orange.

The fruit is a berry with a thick rind. The loculae of the ovary are infiltrated by enlarged cells of the inner layer of the pericarp to give the sacs filled with juice that are so typical of this fruit. The pericarp has in the outer zone many pellucid glands secreting essential oil. Amongst the cultivated citrus fruits the extent of the rind varies from being no more than 5 mm thick to constituting the bulk of the fruit. There is not a clear distinction into exocarp, mesocarp and endocarp, though the white region of the rind (equated with the mesocarp) has been termed the albedo and the yellow outer part (exocarp) the flavedo. Not all the fruits are yellow or orange; many are green. This fruit is called a *hesperidium*. The number of seeds per carpel varies and parthenocarpy occurs. As well as parthenocarpy there is a type of apomixis. Polyembryonic seeds are commonplace, perhaps three or more seedlings arising from a single seed. Only one of these seedlings has resulted from fertilization; the others arise from nucellar tissue and are hence genetically identical to the mother tree. Polyembryony can easily be demonstrated in the hybrid between *Citrus* and *Poncirus trifoliata* L. The latter has trifoliate leaves and this character is dominant. If used as the pollen parent then the trifoliate progeny in a polyembryonic seed is the one which has been produced by fertilization; the others are maternal progeny.

Seeds lose their viability rapidly and in the production of seedling rootstocks seeds should be sown soon after harvesting.

Most of the commercial sweet oranges (q.v.) are grown on sour orange (q.v.) rootstocks because of the latter's resistance to *Phytophthora* spp. especially *P. parasitica* Dastur, responsible for the disease called gummosis.

The most important species are

Citrus aurantifolia (Christm.) Swing.	Lime
C. aurantium L.	Sour orange, Seville orange
C. grandis (L.) Osbeck.	Pummelo, Shaddock
C. limon (L.) Burn. f.	Lemon
C. medica L.	Citron
C. paradisi Macf.	Grapefruit
C. reticulata Blanco	Tangerine, mandarin orange
C. sinensis (L.) Osbeck	Sweet orange

The main distinguishing features of those species entering world trade are

Lime: Thorns pronounced, leaves small, 4–8 cm long, fruits ovoid, small, green, peel thick, adherent, flesh acid, usually 9–12 segments (loculi).

Sour orange: Thorns slender, leaves 10 cm long, fruits subglobose and medium sized, peel orange rough, well endowed with glands (the subsp. *bergamia* (Risso and Poit.) yields bergamot oil), adherent, flesh bitter and acid, 10–12 segments incompletely fused at centre.

Lemon: Thorns stout, leaves ovate crenulate, fruit ovoid with distinct terminal protuberance, peel lemon-yellow, smooth to slightly rough, thick and adherent, flesh acid, segments 8–10.

Grapefruit: Probably of recent origin in the West Indies then spreading to main citrus-growing regions. Not noticeably thorny, leaves broadly winged and blade margins sinuous, fruit large globose, 15 cm diameter, peel greenish to pale yellow, moderately thick adherent, flesh sharp weakly acid, sometimes pinkish (the best varieties), segments 12–14.

Tangerine: Small tree occasionally free of spines, leaves small narrow almost lanceolate up to 8 cm long with crenate margins, fruit globose but with flattened polar regions, peel orange, thin, easily removed, segments 10–15.

Sweet orange: Spines stout, young twigs with characteristic longitudinal ridges, leaves ovate to 15 cm long, fruits subglobose with moderately thick smooth peel, orange when grown in subtropics but greenish at maturity when grown in the tropics, adherent to the mildly acid flesh, carpel segments fused at centre, 10–14 segments.

Pineapple

The pineapple is the other major tropical fruit entering world trade. This trade depends largely on canned rather than fresh fruit, though with more rapid transport and the appearance of fresh 'pines' in the markets, more people now appreciate that the two products are quite distinct. This fruit is technically an infructescence since it consists of a fleshy axis in which are embedded the individual fruits of many flowers.

The plant belongs to the monocotyledonous family Bromeliaceae which is centred in tropical America and the Caribbean Basin though one species is reported from tropical West Africa. There are possibly 46 genera and 1700 species in the Bromeliaceae.

All species exhibit a degree of xeromorphy and many are epiphytic. Stem development is insignificant, and a rosette of tough linear leaves is produced. These leaves may have spiny margins and in a great many the leaf base is urnicate. The rosette composed of such leaves is like a large cup which collects water and debris. Most often the roots produced are for anchoring the plant to its substrate and the necessary water and nutrients are absorbed from the water and debris contained within the leaf bases. In the basal region of the leaves there are special peltate hairs which function as absorptive structures. The liquid in the rosette supports a whole specialized animal and plant community, there being mosquito larvae, exclusive to this ecological niche, and certain *Utricularia* spp. (Bladderworts) a type of insectivorous plant, are also found in the medium.

The flowers are carried in a terminal inflorescence which may at first be on a truncated axis so that it appears almost sessile in the centre of the rosette. Later the inflorescence may be carried on a short stem to be raised clear of the rosette. Individual flowers are subtended by a pronounced bract, often brilliantly coloured. These flowers are trimerous with the outer whorls alike; sometimes the perianth members fuse. There are six stamens in two whorls, then the ovary of three fused carpels. The family has members which have superior ovaries, but all types of arrangement are found. The ovary is trilocular, each loculus with many ovules arranged parietally. A single style with a tripartite stigma surmounts the ovary.

Pineapple is unusual in that it is a terrestrial species. The cultivated plant, *Ananas comosus* (L.) Merr. is not found as a native wild species, but related species occur in tropical South America and it is likely that the present *A. comosus* was selected from a wild growing population by the Indians. It was carried with them in their travels and it has become feral in some districts.

The plant is a long-lived perennial but is treated as a short-lived type, only two or three crops being taken. Seedlings have a small radicle which soon dies, there being an abundant production of adventitious roots from the base of the stem. The roots either enter the soil or remain above ground and grow in the axils of the leaves.

A short thick stem, up to 25 cm long, is produced and this carries a rosette of long (to 1 m), lanceolate leaves, with pronounced marginal spines, though in the 'Kew' pine and some of its progeny the spines are reduced. The leaf bases are clasping and although the rosette does not form a pronounced basin, rainfall and dew can collect within the leaf bases. The leaves exhibit pronounced xeromorphy having sunken stomata, water storage tissue, and exfoliation of abaxial trichomes. They are also markedly sclerophyllous.

The plant branches at the base to produce a set of rattoons, and also gives rise to suckers from the subterranean part of the axis. Branching in the upper stem can also occur.

The plant is a quantitative short-day plant and flowering can be promoted by an application of auxin. The inflorescence is formed by the upper axis elongating and bearing a large number of axillary flowers. Usually the terminal portion is non-floriferous and the flower-bearing region is crowned by a rosette of small broad leaves unlike the long leaves of the basal rosette (Fig. 43). Each flower is subtended by a scaly bract and is trimerous with three fleshy outer

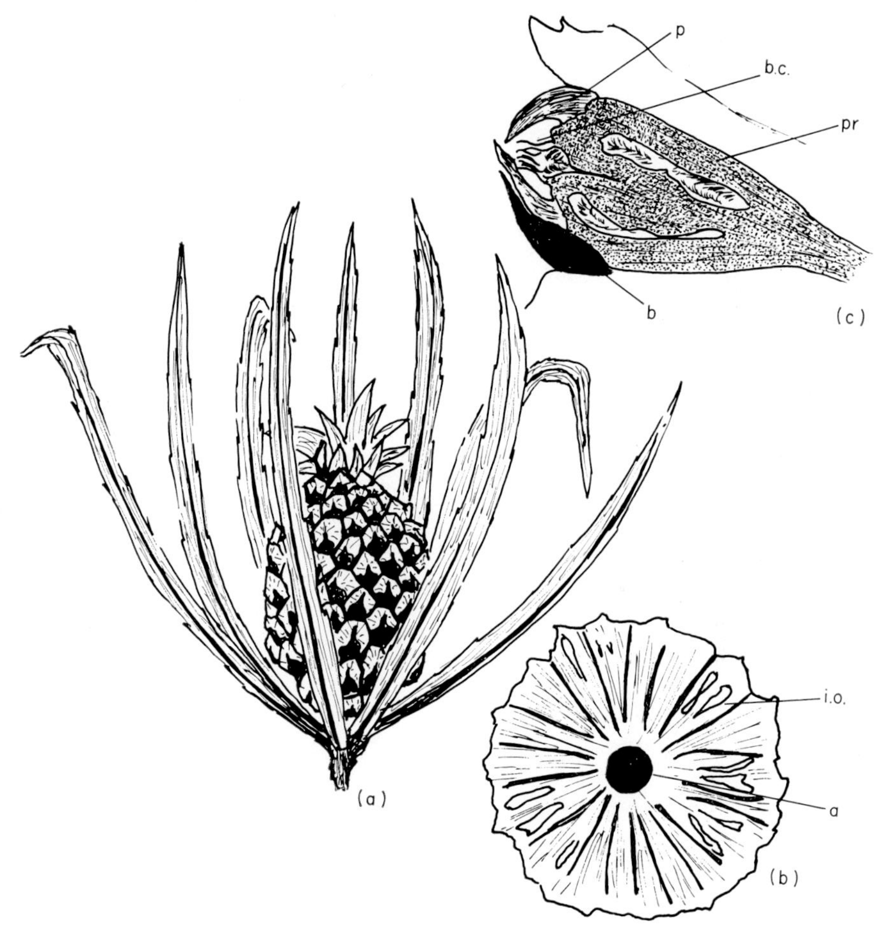

Fig. 43. Pineapple. (a) Whole plant. (b) a, cross-section of syncarp axis; i.o., inferior ovary. (c) longitudinal section of single fruit; b, bract; p, swollen perianth member; pr, swollen pericarp; b.c., blossom cup.

perianth members and three inner, more membranous ligulate tepals which are loosely joined. The six short stamens are in two whorls and there is an inferior ovary of three carpels with a single style and three stigmas. Most cultivated varieties are seedless but if seeded there are up to 20 ovules which can develop further.

At maturity the fleshy outer perianth members enlarge to become the blossom cup, within which are all the other withered flower parts. The ovary wall has likewise become fleshy and with the fusion of these and the outer tepals a large syncarp results. The central axis is distinct and since it is more fibrous it is desirable to have cultivars in which there has been reduction of the axis diameter.

The crown rosette continues to grow during fruit maturity but growth ceases and the crown becomes dormant when the syncarp is ripe.

Normally a planting crop and one or two ratoon crops are taken then the plantation is grubbed. New plantings are made from young ratoons or from axillary branches (slips) arising from below the inflorescence.

The suckers and crowns may be used for propagation but establishment, particularly of the latter, takes longer than in the case of ratoons or slips.

Other shoots produced on other parts of the axis can also be used to propagate the plant.

CHAPTER 11

Some Beverages: Coffee, Tea and Cocoa

Man has been aware of the stimulant properties of some parts of many different plants for probably as long as he has existed. Even in the absence of a specific stimulant, the action of sucking or chewing a straw often results in the obtaining of a psychological satisfaction.

Those plant parts which offer stimulation or satisfaction are either chewed whole or made into an infusion. In the developed countries tobacco (q.v.) was at one time chewed but now it is smoked and the habit of chewing, and its associated expectoration, is considered unsociable. 'Civilized' man has, by convention, accepted that the preparation of an infusion carries with it a cachet of respectability. The manner in which the infusion is done sets a man apart. To cater for this need there is a large world trade in such commodities, nearly 5 000 000 tonnes of coffee, tea and cocoa being shipped annually.

Coffee

Coffee is the most important of this group of plants with an annual world trade involving 3 000 000 tons and a production which is higher. In some years substantially larger tonnages are produced, but careful regulation of the surplus ensures the maintenance of a market price that gives the producers an adequate profit margin.

Coffee is a member of the family Rubiaceae which is one of the larger dicotyledonous families, listed as having 450–500 genera and 6000–7000 species. Most members are trees or subshrubs of the tropics but there are herbaceous species in the temperate zones. In Great Britain the main representative is *Galium* with *Rubia tinctorum* L. less frequent but at one time the species was grown widely for the red dye alizarin (madder). As well as *Coffea*, the genus *Cinchona* is grown for its bark from which quinine is extracted.

Leaves are opposite and decussate with stipules which may be variously modified. In the herbaceous species the stipules may be leaf-like and divided so giving the impression of a whorl of leaves but in others the stipules may be united to give a sheath or are caducous. The leaves are simple, usually entire.

Flowers are carried in axillary cymes, rarely solitary and terminal. The cymes are dichasial. The flowers are regular, hermaphrodite (though can be diclinous) and are either tetramerous or pentamerous. The corolla forms a tube but the sepals are free and there are as many stamens as petals. There is an inferior ovary of two carpels (other arrangements occur). The bilocular ovary may have many ovules per loculus, or only one, and on this feature the family is divided into the subfamilies Cinchonoideae (∞ ovules) and the Rubioideae (Coffeoideae) with a single ovule per loculus.

Coffee is obtained from a number of species belonging to the genus *Coffea* (Fig. 44). This is a large variable genus with possibly as many as 100 species but most taxonomists agree that there are probably around 60 good species. All are distributed in the Old World with the greatest number of species in tropical Africa. Two species, *Coffea arabica* L. (Arabian coffee) and *C. canephora* Pume ex Frochner (Robusta coffee) are in large scale production. They are very similar, differing in vigour, the latter having large leaves 20 cm long or longer in comparison with the former with leaves just over half that length.

C. arabica is a small tree or shrub to 5 m when left unattended but in culture it is pruned to allow easy harvesting of fruits. Coffee is characterized by having two types of branch, a vegetative upright stem system and a more or less horizontal flowering branch. Normally only the main axis is vegetative and the flowering branches which develop from the upper of the two buds present in the axils grow out to give the framework of the tree. If the apical dominance is removed by damage to or pruning of the main stem the lower bud grows out and is pronouncedly negatively geotropic.

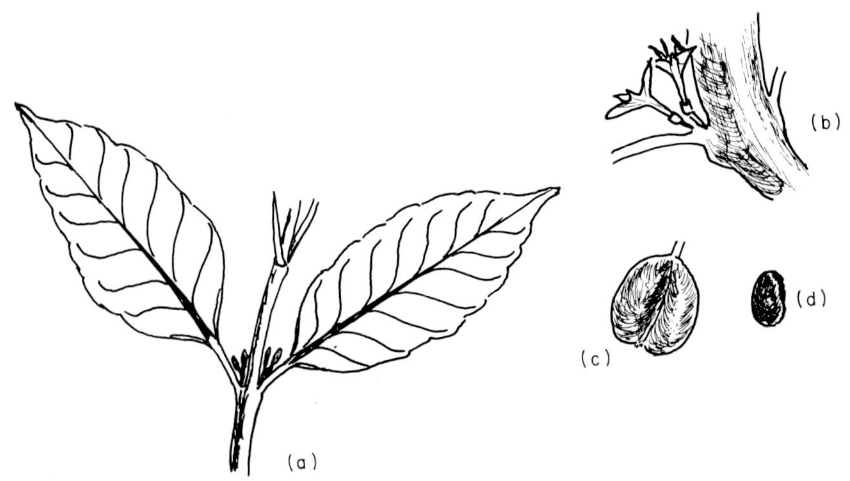

Fig. 44. Coffee. (a) Branch. (b) Inflorescence. (c) Fruit. (d) 'Bean'.

The leaves are shortly stalked and have small stipules lying on the stem between the petioles (interpetiolar). There is a pronounced robust acuminate tip and though the adaxial surface is smooth the midrib region is seen to have a row of bumps corresponding to the vertices made by the lateral vein with the main vein. These bumps are in fact due to *domatia* located on the lower surface of the leaf. It is probable that some small insect or mite makes its habitation in the domatium and the plant may gain some benefit from the association. It is of interest to note that myrmecophily occurs in this family and in the genus *Psychotria* there is a peculiar association between a bacterium in the leaf and the plant. Claims have been made that in *Psychotria* there is nitrogen fixation which is advantageous to the plant but there is a possibility that the plant obtains not a source of metabolic nitrogen but a supply of a plant growth hormone (a cytokinin).

The flowers are very fragrant and are in axillary cymes of up to twenty. Flower production is continuous but flower bud opening is cyclical and depends on a temperature shock and a supply of water. Both are provided by rain which will lower the bud temperature and also provide at the bud a supply of the necessary water. Flowers are about 1 cm across and are pentamerous. The calyx is weakly tubular but the corolla has a pronounced basal tube. The isomerous stamens are epipetalous and there is a pronounced inferior ovary of two carpels. The style is long and has a bifid stigma. There is one ovule per carpel. The fruit is drupaceous containing two seeds each with a large endosperm.

Flower buds which are not provided with sufficient water will open to give a depauperate flower which has vestigial anthers. These are called star flowers and are often wholly green.

At maturity the fruit is a rich deep red and the fleshy part is partly removed by fermentation. The endocarp is papery and forms the parchment which is silvery, and when this is removed the seeds are the coffee beans of commerce. Before being ground the beans are roasted to varying degrees to give the different types of coffee for particular palates.

Coffee is a plant of the understorey of tropical forests. It cannot tolerate wide extremes of temperature either diurnally or seasonally, and its growth is only successful in specific regions of the world. In hotter zones it can be grown in shade, but if water is limited the shade trees compete for the available moisture and flower opening in coffee is prevented. Some systems of coffee growing plant four plants to a hill and it is possible that some mutual shading is achieved.

Tea

While tea has been used in the Far East for millennia, its use in Europe has only been widespread since the 18th century—a little after the introduction of coffee. Its current use is not as widespread as coffee but it still remains the most popular non-alcoholic beverage in Great Britain.

Tea is derived from the terminal shoots of the small shrub *Camellia sinensis* L., O. Kuntze. This plant has been variously named but it is agreed that it is a member of the genus *Camellia* and is not a distinct genus *Thea*. It belongs to the family Theaceae (Ternstroemiaceae). This is a small family of 35 genera and 600 species, mostly woody and of the tropics and subtropics. The leaves are simple, alternate and leathery. Flowers are solitary (Fig. 45) and of irregular arrangement and some of the parts may be arranged spirally. There are five to seven persistent sepals, five petals, or a multiple of five; stamens are indefinite but the superior ovary has up to ten loculae with as many styles as chambers. The loculae each may have many ovules. The fruit is a capsule or drupe.

Tea itself is a small tree but is trained as a bush. Branching is encouraged to provide many stems which can give the tips that are collected. Shoot growth and leaf production are managed to give a succession of flushes which can be picked, and major pruning and shaping are confined to the dormant season,

Fig. 45. Tea. Terminal region of branch with flower in lower axil.

with severe pruning being carried out in alternate years or at longer durations depending on the vigour of the plant and the locality.

The leaves are alternate, evergreen, ovate, sharply pointed and with a pronounced serrate margin, and may reach 30 cm. Hairs are present on the lower surface but the upper surface is glabrous. At flushing the first true leaf that is produced is small and blunt; only subsequently do the leaves assume the normal shape and this is associated with internode extension. After some four leaves are produced shoot growth declines and the shoot enters a dormant phase for the next flush to occur from the axillary bud of the uppermost leaf—a classic case of sympodial development.

The flowers are fairly small, not more than 4 cm across with five to seven sepals, five to seven petals, numerous stamens and the ovary has from three to five styles connate at the base, with free stigmas. There are up to six seeds per chamber in the capsule.

The plant is extremely variable; small, dwarf types are found in China, with robust vigorous forms in Assam. Purseglove states that a *cline*, based on certain morphological features, can be erected between these extreme types. This continuous variation could be treated as clinal but unless correlated with the habitat such a morphocline has limited taxonomic value. Certainly the extremes are distinct and have been given varietal status, viz., *C. sinensis* var. *sinensis* and *C. sinensis* var. *assamica*.

The young enlarging shoot is plucked before the leaves develop a substantial cuticle. The highest quality teas are produced from the terminal bud and adjacent leaves, and as more leaves are included in the picking so the quality is lowered. The plucked leaves are withered, rolled and fermented to produce black tea but in green tea 'fermentation' is prevented. The 'fermentation' of tea is not a microbial fermentation but an oxidation of polyphenols, contained in quantity in the bud and youngest leaves, by enzymes present in the leaf. The rolling, which involves bruising the leaf, brings the enzymes into contact with the substrate. For rapid fermentation the rolled leaf is broken and aerated, this process aiding access of oxygen to the leaf. After fermenting, the tea is dried (fired) by hot air at between 80 and 95 °C. China tea is produced by firing immediately after withering, or by destroying the enzymes by exposing the plucked leaf to steam.

Many grades of tea can be produced according to which picking, withering, rolling and fermentation procedures are adopted.

Cocoa

Cocoa is not so much an infusion as a suspension of the finely ground seed of *Theobroma cacao* L. This tropical tree is typical of the family Sterculiaceae; its seeds were introduced into Europe from S. America in the 17th century and the drinking of chocolate, a beverage prepared from cocoa beans, vanilla essence and sugar, became as popular as coffee.

The Sterculiaceae is a family of 70 genera and 1000 species, mainly distributed in the tropics. They are nearly all small trees or shrubs with simple stipulate leaves. The flowers are carried in cymes and, in the case of *Theobroma*, the inflorescences are *cauliflorous*, i.e. they arise directly on the trunk and are not associated with either the apices or axils of the plant (Fig. 46a). In the flower there are five sepals united into a tube, five petals when a corolla is present, ten stamens in two whorls, again, like the calyx, united into a tube, at least at the base, and usually a superior ovary of five fused carpels (Fig. 46b).

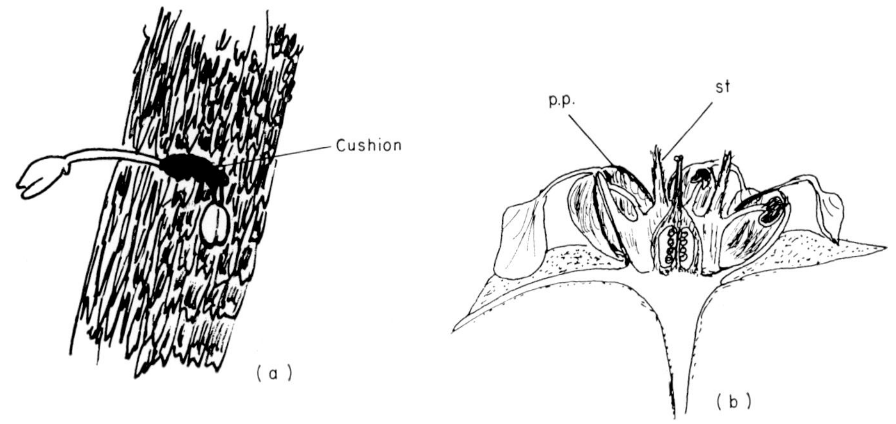

Fig. 46. Cocoa. (a) Flower cushion with two flower buds. (b) Half-flower: p.p., pouched petal; st, staminode.

T. cacao occurs naturally as a small understorey tree some 6–8 m high. Like a number of other tropical small trees it shows two types of branching (dimorphic branching) (cf. *Coffea*). In cocoa the plageotropic branches arise together at what seems to be a whorl at a nodal region which is swollen. This branch system is called a *jorquette*. The branches arise not as axillaries but as a result of the terminal meristem dividing, and each of the separate components goes on to produce a plagiotropic shoot. Further upward growth of the plant is accomplished by the activation of an axillary bud below the jorquette to produce an upright, negatively geotropic shoot called a *chupon*. A jorquette can develop on the chupon. The pattern is repeated.

The large simple leaves with lanceolate stipules are arranged alternately, but with a phyllotaxy of $\frac{3}{8}$ on a chupon and $\frac{1}{2}$ on the branches of the jorquette (fan branches). The stipules are caducous. The leaves may be up to 60 cm long and are elliptic to obovate with a prominent main vein and well-developed pinnate venation with the ends of the lateral veins closed.

The inflorescences are produced on the old wood from the base of the first chupon, i.e. the sapling stem, all the way up to the older jorquettes. Young wood never carries inflorescences. The inflorescence arises on a cushion and these cushions are perennial so that when fruits are removed it is essential to ensure that the cushion is not damaged. While it was remarked earlier that the inflorescence appears not to be associated with an axil, this is not strictly true. The cushions arise as dwarf shoots, give rise to the inflorescences, and in subsequent development there is the build-up of the cushion.

The flowers are small and pinkish with the cyme carrying perhaps 20 flowers. The flower stalks and the bracts are slightly pubescent. Each flower is regular, hermaphrodite and pentamerous, there being a calyx of five valvate sepals. The corolla is unusual in that the five petals are each pouched in the centre and broaden to a spathulate tip. There are two whorls in the androecium, an outer staminodal whorl of five members, petaloid, pointed and with ciliate margins, and an inner whorl of five functional stamens bent outwards and with the anthers contained in the pouch of the petal. All the members of the androecium have the basal regions fused to a tube. The superior ovary is of five fused carpels and has a single style with five adherent stigmas. There are numerous ovules and the placentation is mixed, axile at the base of the ovary and parietal above.

The fruit is large in comparison to the flower and not every flower of a cushion gives rise to a fruit. At maturity the large fruit, technically a drupe but referred to as a pod, may be up to 30 cm long, but the shape varies from nearly spherical to elliptical. It is usually pointed but can be blunt. The seeds, as many as 60 per pod, are ovoid, up to 4 cm long and contain a large embryo with ruminate cotyledons, a papery endosperm, and have a leathery testa. The seed is rich in fat ($>50\%$) and this is removed as cocoa butter.

The seeds are surrounded by mucilage and on harvesting the pods, which have a leathery exocarp and a pulpy fibrous mesocarp, the seeds, along with the mucilage and remnants of mesocarp, are fermented. Fermentation takes up to 7 days and may be termed sweating. Special fermentation boxes, not more than 1 m deep and with well-spaced slats, are often used to permit adequate aeration during the second stage when *Acetobacter* replaces yeasts as the main fermenting agent. During fermentation the beans undergo a change of composition. Pigment is lost as are astringent principles whilst the development of the typical cocoa flavour takes place.

On completion of fermentation the beans are dried by exposure to the sun. Barbecues with sliding covers to shield from rain and dew are often used. After drying the beans are polished.

Theobroma shows self-incompatibility. Some individuals do not exhibit this phenomenon. This feature is important—not in the production of trees, since most good planting material is propagated as clones by cuttings, but in the production of fruits. If fertilization does not take place the flowers are shed.

Another important factor in the consistent production of fruit is pollination and here it has only recently been found that the main pollinator is a midge.

The midges are attracted to the staminodes and walk about the flower to the anthers in the petal pouches. Transfer of pollen is inefficient and it has been estimated that in some regions as few as one in 500 flowers ever set fruit.

Cocoa is a native of northern South America and is one of possibly some 20 species of *Theobroma* of that area. Wild *Theobroma* is still harvested in the area and amongst the wild forms there are recognized the two main types of *cacao*: Criollo, with yellow or red pods, deeply furrowed and with a marked pointed apex, and Forastero, with yellow pod, almost smooth, and with a rounded apex. These have hybridized to give the Trinitarios which are very variable. Criollos give the best cocoa, but are low yielding, and have originated in the region from southern Mexico to Venezuela. The Forastero types are more southernly and are well exemplified in the Amazonian Basin. There is much variation in both these types and this has led to some difficulty with the taxonomy, but the following is acceptable.

T. cacao L. subsp. *cacao* f. *cacao*. Criollo—Central American.

T. cacao L. subsp. *cacao* f. *pentagonum* (Bern.) Cuatr. Criollo type—Central American, known as Alligator cocoa, and is a cultivated form.

T. cacao L. subsp. *cacao* f. *leiocarpum* (Bern.) Ducke. Criollo type known only in cultivation.

T. cacao L. subsp. *cacao* f. *lacandonense* Cuatr. A unique form with a vine-like habit.

T. cacao L. subsp. *sphaerocarpum* (Chev.) Cuatr. Forastero type.

Cola is a masticatory belonging to the family.

CHAPTER 12

The Euphorbiaceae: Rubber

This is a large and very diverse family. It includes some 290 genera and 7500 species; herbs, shrubs and trees are represented. While the distribution is mainly in tropical Africa and South America, many of the herbaceous species are found in the temperate zones of both the North and South hemispheres. The family presents some taxonomic problems but it is possible to recognize some natural groups within what is otherwise an unnatural assemblage.

A great variety of habit is encountered, from xerophytes with reduced leaves to climbing lianes. Leaf arrangement is usually alternate, but opposite leaves occur and in some species phyllatoxy can change within the plant from alternating in the basal regions to opposite in the upper branches. Stipules are present but may be modified, venation marked with leaves which may be deeply lobed.

The flowers are small, unisexual, often reduced and arranged in cymose clusters which may become complex e.g. in *Euphorbia* itself. The outer floral parts are represented by a perianth, usually five in number, but both a calyx and corolla may be present or these structures may be absent. The stamens vary from one to indefinite and may have branched filaments (e.g. *Ricinus* q.v.). The ovary is three-celled, consists of three fused superior carpels and is surmounted by a prominent two-, sometimes three-, lobed stigma. The gynoecium is the most constant character of the family.

Many species contain a milky latex and most species are poisonous, the seeds usually being the most poisonous part of the plant. The seeds are formed from pendulous anatropous ovules, two per loculus, and these ovules have a pronounced raphe. At maturity a caruncle covers the micropyle and adjacent areas. Most seeds are rich in oil.

The family has given us three distinct categories of crop:

1. Tuberous-rooted Cassava
2. The oil crops Tung and Castor
3. Rubber

Cassava, manioc, or tapioca, is now grown widely in the tropics where it is one of the more important starch crops. It is accepted that the area of natural distribution is in Central America with a secondary centre in Brazil. During the 18th century it was transported to the Old World and today Africa produces the largest quantities. Except for processed starch, tapioca, it is not now an important item in world trade.

Cassava belongs to the species *Manihot esculenta* Crantz. but this species has at various times been redefined to separate the so-called bitter cassava from sweet cassava. As remarked, most Euphorbiaceae are poisonous and cassava tubers contain a cyanogenetic glycoside. This is removed by fermentation and soaking in water, a process which can take a week, and then the tuber is pounded to release the starch. The bitter cassavas contain much glycoside distributed throughout the tuber but sweet cassavas have the glycoside only in the outer layers. If the phelloderm is removed the central tissues can be prepared without prior fermentation. However, the distribution of glycoside in the tuber is due as much to environmental factors and the stage of growth, as it is to genotype so it is preferred to assign these forms to the same species.

Cassava is a short-lived perennial shrub which is persistent by virtue of possessing swollen adventitious roots. These tubers are cylindrical, tapering and may be up to 1 m long. The central region of the tuber is rich in large starch grains. The poisonous principle may be destroyed by boiling or roasting but in the preparation of the starch, fermentation is preferred to rid the tuber of toxin. Latex is also present and this may in itself be bitter.

The leaves, alternately arranged, are stipulate and with petioles to 30 cm. The blade, shorter than the petiole, is variously palmately lobed but usually with five to seven lobes deeply cut and with a pronounced tip. Sometimes the leaf is slightly hairy on the abaxial surface.

The primary inflorescence consists of axillary racemes with the flowers in clusters, both male and female flowers in the basal clusters, with the upper ones usually consisting only of male flowers. Male flowers are pedicellate and have five sepals and ten stamens, five long and five short, in two whorls. The sepals are coloured pale yellow, tinged with red and are joined at the base. Female flowers are larger, on a pronounced pedicel, and with the calyx members separate. There is a prominent tricarpellary ovary surmounted by a style carrying a robust tripartite stigma, each part much lobed. The fruit is a capsule containing three seeds each about 12 mm long, mottled with dark blotches and a pronounced caruncle.

Tung and castor oils are much prized drying oils, though at one time the latter was more important as a medicinal oil. With proper processing, a lubricating oil able to withstand higher temperatures than mineral oils can be obtained from castor oil, and there is a substantial demand for this. Another member of the Euphorbiaceae produces a seed oil with more pronounced medicinal properties than castor oil, croton oil (from *Croton tiglium* L.) but this is not a regular pharmaceutical product.

Tung oil is obtained from the small tree *Aleurites montana* (Lour.) Wils, or from the related species *A. fordii* Hemsl. *A. montana* cannot grow outside the tropical zone whereas *A. fordii*, a native of West-central China, is suited to cooler regions; indeed it is unsuccessful as a crop if grown in a tropical climate.

The more important of these species is *A. montana* which is now grown in Central Africa, but there has not been an increase in the acreage grown in the last few years. It is a tree reaching to 20 m with spirally arranged leaves which may be entire or three- to five-lobed. The petioles are longer than the laminas which are broad. The leaf is characterized by possessing, at the junction of the lamina and petiole, a pair of stalked green extra-floral nectaries.

Flowers are carried variously with regard to distribution of sexes, some trees being predominantly male and others female, and yet others having equal distribution of male and female inflorescences. These inflorescences, basically racemose, are borne on new season's wood and the plant is strongly protogynous though the male inflorescences appear first. In the management of tung it is essential to have good pollination, and to obtain this a few predominantly male trees, which have an extended period of pollen production, should be planted as pollinators. This is easy to arrange since commercial plantings are done with cloned material budded on to seedling rootstocks.

The male inflorescences have up to 200 flowers. Each of these flowers has a two- to three-lobed calyx and five petals enclosing eight to twenty stamens. Between the petals and the first of the three whorls of stamens are five glands. Female flowers are larger and are in denser inflorescences of about 20 flowers. These flowers have the same outer parts and the ovary is three- to five-celled, eventually having a single seed in each cell. A style with two robust branches terminating in large stigmas surmounts the ovary. The fruits is drupaceous and the seeds, up to 3 cm long, have a white endosperm.

The seeds themselves have an oil content which can be as high as 40% but if they are not harvested separately the oil content of the fruit is only about 15%. It is usual to remove the seeds from the fruits before expressing the oil. A recovery of 75% is usual but the oil seed cake left, which has a high nutritional value on analysis, can only be used as fertilizer because of its toxicity.

Castor oil is produced from the seeds of *Ricinus communis* L. (Fig. 47). Purseglove states that this is a monotypic genus and this view prevails amongst taxonomists. Other specific epithets used such as *gibsoni* and *zanzibarensis* are invalid. The species is very variable and it is easy to appreciate how recognizably distinct genotypes have been afforded specific status by those trading in seeds.

This plant has been grown since antiquity and at various times the oil extracted from the seed has been used as an illuminant, medicinally, or as a lubricating and a drying oil. It grows vigorously as a subshrub and while considered an annual can persist for a number of years in favourable environments. A well-grown plant can be up to 7 m high. The stems are robust and carry spirally arranged leaves which are stipulate, stipules to 3 cm and adnate to petiole, with petioles shorter than the blade. The blade is large, up to 75 cm in diameter, palmately lobed with from five to eleven divisions reaching half way to base. The segments are serrate, lanceolate, dark above, pale below and may be suffused with red pigment. There are two extra floral nectaries at the junction of the blade and petiole.

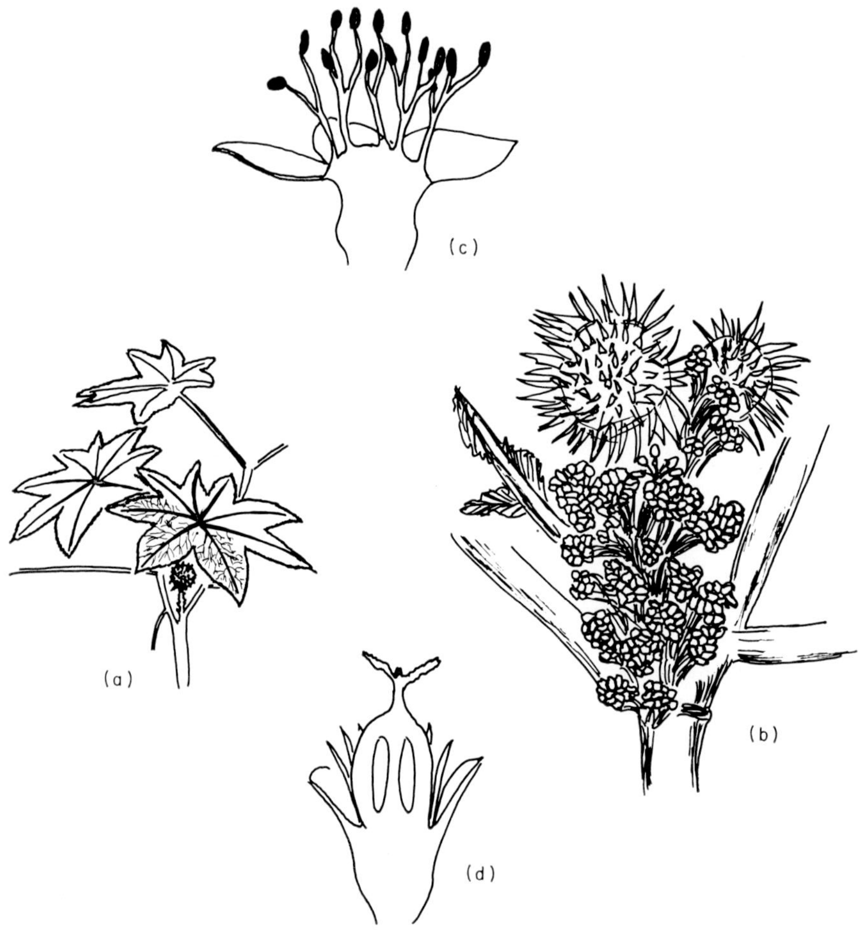

Fig. 47. *Ricinus.* (a) Portion of whole plant. (b) Inflorescence. (c) Male flower. (d) Female flower.

The flowers are carried on many-flowered terminal inflorescences. Male flowers in cymose clusters of from three to sixteen are basal while the female flowers, numbering one to seven in small cymes are borne towards the top of the inflorescence. Female flowers are more robust, on pedicels to 5 mm long; the males are smaller, on larger pedicels. The sepals number three to five and are fused in the female flower. Male flowers contain numerous branched stamens and in the female there is a three-celled ovary with one ovule per cell. The style is short and has three fleshy stigmas.

The seeds vary greatly in colour and size, but all have a light-coloured caruncle. The seed contains the poisonous protein *ricin* which can cause death if ingested and can also result in severe allergy if it is absorbed through cuts in the skin. As well as ricin the seed contains a specific allergen which can be troublesome. Neither of these poisonous principles finds its way to the oil but they are left in the cake obtained after expressing the oil. The meal may be detoxified but unless this is done it can only be used as a fertilizer. As expressed the oil is non-drying but after dehydration the oil can be oxidized at the double bonds to give a varnish which is colourless. This makes it most valuable in paint manufacture.

The remaining member of the family we shall consider is *Hevea brasiliensis* (Willd. ex Adr. de Juss.) Muell.-Ang. from which is obtained Para Rubber. This rubber is now about 99% of the world's natural rubber, and though natural rubber was eclipsed after World War II by the introduction of synthetic rubber, the demand for rubber is so great, and the concern felt about the use of scarce resources ever present, that the acreage planted to *Hevea* continues to increase.

Hevea is considered to contain nine species but this number has been as high as twelve. All species are well endowed with latex in all parts of the plant but one species, *H. brasiliensis*, produces a latex which can be coagulated, usually by smoking or addition of ammonia. In the latter an emulsion is formed of the coagulum. *H. brasiliensis* can be harvested continuously, for the cut made to extract the latex can give a flow over a few days, due to the latex not coagulating quickly when exposed to air and can then be renewed to permit latex to continue flowing. This *tapping* can be performed frequently and latex flow can be stimulated by treatment with 2,4-D or 2,4,5-T (hormone-like chemicals), or Ethrel.

The latex vessels are found in the secondary phloem; indeed they are modified sieve tubes, and run in a counterclockwise spiral. The greater number are nearest the newly formed phloem. If the tapping cut is made in a clockwise direction and as close to the cambium as possible, more of the active latex tubes are cut and latex flow is maximized. Substantially higher flow rates than average can be obtained by expert tapping.

Hevea is a quick-growing tree seldom taller than 25 m. The leaves are trifoliate with caducous stipules. There are three extra-floral nectaries at the junction of the leaflets with the petiole. The petioles are usually smaller than the leaflets which vary in size. The sharply pointed leaflets are entire, obovate, and shortly stalked.

The inflorescences are axillary on the new growth and have many small male flowers towards the base with a few larger apical female flowers. The flowers with a calyx of five members are sweetly scented. Male flowers have 10 stamens in two whorls, these stamens consisting of sessile anthers on a central column. Female flowers have a three-celled ovary surrounded by a disc. There are three stigmas. The fruit is a three-lobed capsule, very common in this family, each lobe having a single seed. The seeds are large, averaging 3 cm × 2 cm, but are not used in propagation. Most cultivars are proven clones.

Hevea is native to Brazil and before it was taken by the British to S.E. Asia all rubber was obtained from wild growing plants and other species were involved in the production. Not all species can tolerate plantation conditions and it is this property, as much as any special virtues *Hevea* latex might have which has been responsible for the remarkable success of this crop plant. During the Second World War when the main plantation areas were in the hands of the Japanese the Allied Powers sought other plants that might substitute for *Hevea* without success. Indeed, it was this failure that resulted in the expansion of the synthetic rubber industry.

CHAPTER 13

The Umbelliferae: Carrots and Allies

The members of this family have such a characteristic habit that they are readily recognized. There are perhaps some 300 genera with 3000 species in it and except for those species which belong to the subfamily Hydrocotyloideae all possess compound leaves, and at flowering have robust stems with hollow internodes. At the nodes the leaves have a sheathing base encircling the stem. It is the inflorescence which is the most distinctive part of the plant, and this is best seen in the members of the largest subfamily the Apioideae, with 260 genera. The third subfamily is the Saniculoideae which is small with nine genera, and though there are interesting species in both it and the Hydrocotyloideae, all the crop species have been derived from ancestral types which are members of the Apioideae.

The inflorescence found in this family is the umbel or the compound umbel and so typical are these that the family has been given the name Umbelliferae, the umbel bearers. It is worth remarking that in some genera of the minor subfamilies the inflorescence is basically cymose, not racemose, though the overall form of the inflorescence is still that of the umbel. These cymose umbels can be distinguished by the order in which the flowers open, this being centrifugal and not centripetal.

Most species are robust herbs, the vegetative stage consisting of a telescoped stem giving a rosette of leaves. At flowering the plant becomes caulescent and carries terminal inflorescences on the main axis and any laterals which have been produced. Annuals, biennials and perennials are found and since the plants are mainly distributed in the temperate regions it is not surprising to find that they require to be exposed to low temperature and the appropriate daylength in order to flower. The biennial and perennial species often develop substantial rootstocks as overwintering organs, and it is from such types that some of the crop species have been selected.

The leaves are often much dissected, nearly always compoundly pinnate. Once-, twice- and thrice-pinnately compound leaves are encountered and the degree of dissection of the leaf is a valuable diagnostic feature. Many of the plants smell characteristically when bruised and often the smell can be a useful aid to identification. The petioles can be fleshy and as remarked have a sheathing base. The stems may or may not be ridged and are either glabrous or hairy.

The inflorescence is made up of small flowers, generally actinomorphic, except for the outermost members which may be zygomorphic, so conferring upon the whole inflorescence a form which may be likened to a single large flower (Fig. 48a). This particular phenomenon, where the peripheral flowers of an inflorescence are larger than the central ones, we have met already in the genus *Iberis* (Cruciferae), but it is seen most highly developed in the Compositae. The simple umbel is the less common type of inflorescence, and the compound umbel in which a series of umbellules is carried in an umbellate head is that most often met in the Apioideae. The umbels may be subtended by bracts and the umbellules by bracteoles, often of characteristic shape and of considerable diagnostic value.

The individual flower is composed of five small sepals, five petals, usually

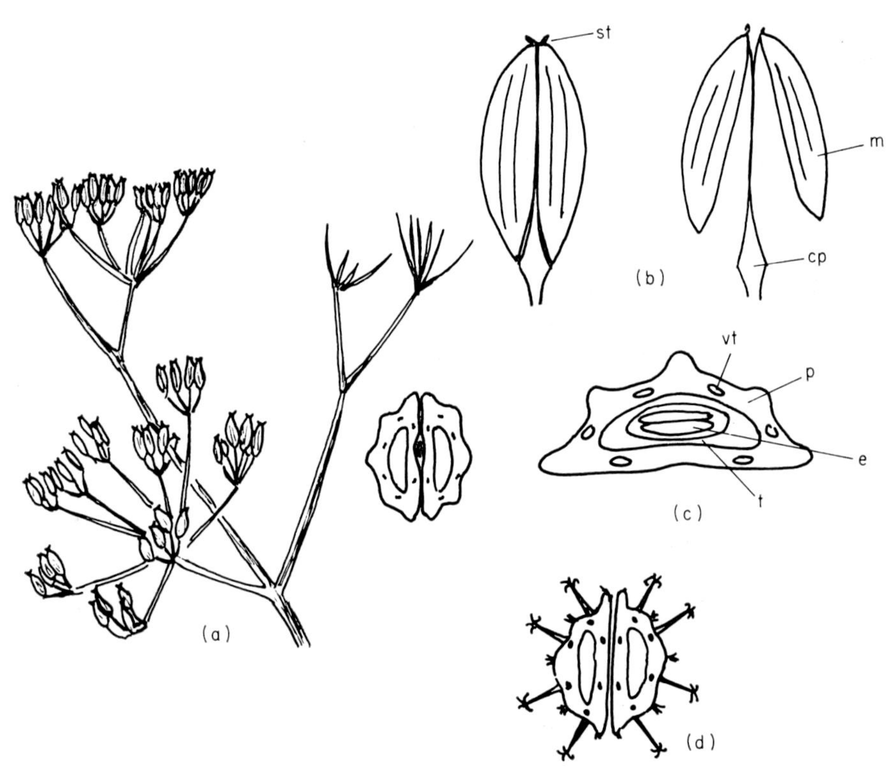

Fig. 48. (a) Fruiting head of *Myrrhis odorata* L. showing development of compound umbel. (b) Diagram of typical umbelliferous fruit: cp, carpophore; st, stylopodium. (c) Transverse section of fruit and individual mericarp: p, pericarp wall; vt, vitelline gland; e, embryo. (d) Cross-section of carrot fruit showing development of spines from secondary ridges.

white but may be yellow, alternating with the sepals, and five stamens alternating with the petals and two inferior carpels. The carpels are arranged in a most interesting way and have a distinctive morphology. The receptacle terminates as a fine stalk, the *carpophore,* which runs between the two carpels. The tip of the carpophore is associated with the top of the ovary to form a disc which secretes nectar, the *stylopodium.* Above the stylopodium there is a short stigma which ends in two distinct stigmas.

Each carpel wall is flattened along that part in contact with the carpophore, but is roughly semicircular in outline on its outer face. This outer face is not smooth having along its length three ridges. These three and the two laterals formed where the carpel leaves the carpophore constitute the primary ridges. Between the primary ridges of the outer face there sometimes develop four additional secondary ridges. On the ridges there may arise spines or hooks. Within the carpellary wall of most of the Apioideae there are six glandular canals, two in the inner face and four, lying between the primary ridges, on the outer face. These are the *vittae,* or *vitelline glands,* which secrete essential oils. Many Umbelliferae are grown for these oils, e.g. coriander, cumin and caraway. Each carpel has a single ovule and at maturity the single seed has a small oily endosperm. The cotyledons are also rich in oil. Sometimes the seed is fused to the pericarp (cf. Gramineae). The oil within the seed is rich in different flavourings, but even if it were an edible oil it is not practicable to extract the seed oils without obtaining the vitelline oil at the same time. Many seed oils from this family are used in cosmetics, as industrial flavours and in the processed food industry.

The fruit is the dispersal unit and while it is schizocarpic, splitting longitudinally up the carpophore to release the individual fruits, termed *mericarps,* it is often found that the separation to mericarps does not occur in commerial 'seed'. This is not as troublesome as in the Chenopodiaceae but it is nonetheless a feature to be aware of in the production of seed.

The individual flower is markedly protandrous, all pollen being shed before the stigmas become receptive. This mechanism is not effective in preventing self-fertilization because within an umbel there is always a range of maturities of flowers, so while self-pollination within a flower does not happen, the genetic equivalent, self-pollination within an inflorescence, is a likely possibility. The pollinating insects are small flies and they tend to work over a single inflorescence rather than go from plant to plant. The spectrum of flower maturities is reflected in the spectrum of ripeness of fruits seen on a plant, and unless left until the youngest fruit is ripe, with the possible loss of over-ripe fruits, some 'seed' will always be immature at harvest. This feature of a range of maturities of fruit might be a contributory factor to the irregular germination met in the Umbelliferous crops. However, it is not the main reason for the long protracted germination; that lies more with the material secreted in the vittae. Most of the Umbelliferous seed oils contain potent germination inhibitors.

Carrot and parsnip are the most important crops, and while they have great potential as producers of digestible material for cattle and sheep, the difficulties of removing long heavy rootstocks from soils which are wet, and in the winter, have resulted in these plants being grown more for human consumption. In warmer climates, especially in the Mediterranean basin, the species which give essential oils are cropped.

The Umbelliferous crops are not the only important members of the family. Most of the species are poisonous to some extent, and fatalities amongst stock and man are not uncommon. Perhaps the most dangerous species are those which can be mistaken for the crop species, e.g. fool's parsley, which fortunately is not highly toxic, or water dropwort hemlock, which can be mistaken for artichokes. The tuberous roots of this last species are also called 'dead man's fingers' and this is a most apt description.

Carrot, *Daucus carota* L., is a plant found growing wild throughout north-west Europe, especially on sandy soils. It is possible in as few as five generations to make selections from native material and produce a plant with recognizable crop characteristics. The plant grows either as an annual or biennial, the shorter lived forms being best adapted to forcing and for horticultural purposes, and the longer-lived ones providing material which leads to the production of those cultivars developed for cattle feeding.

After a long period for germination, up to five weeks, the seedling produces two strap-shaped cotyledons, and shortly thereafter the first foliage leaf. This leaf is multipinnate, but not as complex as the adult leaves. Almost as soon as this first leaf is expanding secondary thickening starts in the hypocotyl and root. The stem does not elongate but there is the prolific production of leaves which form a compact rosette. In modern varieties these leaves are upright.

The secondary thickening which takes place is normal. The stele of the root is diarch, and a cambium is laid down between the primary phloem and the primary xylem. The two isolated fragments of cambium are joined together as a result of development of more cambium in the region of the protoxylem. This cambium soon becomes circular in outline in transverse view, and produces secondary xylem towards the centre and secondary phloem to the outside. The cortex is ruptured but the developing rootstock is protected by a periderm which is formed from a cork cambium which arises in the pericycle. There is equal production of secondary xylem and secondary phloem, and in the mature root these zones are easily recognized because of the lighter colour of the secondary xylem. Carrots can be white, yellow or orange. The highest yields are obtained from the white roots, but the orange ones are of the highest quality. White carrots have a substantial part of the storage structure derived from the hypocotyl, and when mature may have a crown that can be 15 cm above ground. Roots of various shape are grown for different markets. Stump-rooted varieties are favoured for forcing, and the long tapering roots are preferred as stock feed.

Parsnip, *Peucedanum sativum* Benth., is a plant with the same habit as carrot. Again there are annual and biennial forms the latter being the higher yielders.

The 'seeds' of parsnip are flat, and like those of carrot have a protracted and irregular germination. The cotyledons are larger than those of *Daucus*, and the first-formed leaves are simple or with a few small pinnae. The plant forms a rosette and the adult leaves are once-pinnate with two to five pairs of pinnae which each can be as much as 7.5 cm long. The petiole is robust.

Parsnip inflorescences are compound umbels without bracts or bracteoles. The flower is as described for the family, but the lateral ridges of the carpel are prolonged into wings.

The storage 'root' of parsnip is like that of carrot but at maturity it is white and has a higher dry matter content. Unlike carrot the main food reserve of the 'root' is starch.

Amongst the salad crops celery is important, both for use directly in salads and as a flavouring agent. This plant, *Apium graveolens* L. var. *dulce* (Mill.) D.C. and also the var. *rapaceum* (Mill.) D.C., the celeriac, are widely grown but require rich deep organic soils with plentiful supplies of water in order to yield adequately. In celery the petioles are very robust whilst in celeriac the basal part of the petiole surrounds a swollen stem.

The leaves of *A. graveolens* are simply-pinnate with deltoid to rhomboidal leaflets, often lobed and serrate. Lower leaves are long-stalked but the upper leaves are sessile and ternate. The inflorescences are compound umbels without bracteoles.

The agricultural importance of the Umbelliferae is not confined to these crop species. Many are extremely poisonous and constitute a hazard both to man and to his animals. The poisonous substances may be toxic e.g. coniine, or irritant, e.g. the furanocoumarins, or can cause serious metabolic disturbances. Amongst the most frequently encountered harmful members of this family are

Conium maculatum L.	Hemlock, with red spotted stems and smelling of mice when bruised.
Cicuta virosa L.	Cowbane.
Aethusa cynapium L.	Fool's parsley.
Oenanthe crocata L.	Hemlock water dropwort, dead man's fingers, possessing heavy yellow root tubers.

CHAPTER 14

The Liliaceae: Onions and Allies

The genus *Allium* to which onion belongs, has been included in the family Liliaceae by most angiosperm taxonomists. The order Liliiflorae erected by Engler included 13 families but later treatments giving different weight to the characters used in classification arrived at more families and/or the reallocation of genera amongst the families. Hutchinson placed more emphasis on the inflorescence as a family character allocating to the Amaryllidaceae all genera which have umbellate inflorescences subtended by membranous bracts regardless of the position of the ovary. (In extreme cases the inflorescence may be a single flower subtended by a single bract.) A later treatment of this order renamed Liliales, places the genus *Allium* in the family Alliaceae and creates a number of separate families, each with fewer genera, that were formerly within the Liliaceae or Amaryllidaceae.

These families are very 'natural' and while it is possible to establish other groupings that could be construed as families, the writer does not consider that further separation is justifiable in this treatment.

The Liliaceae is a moderately large family of monocotyledons with 220 genera and 3500 species and their structure is that of the typical classic monocotyledon. Most members are perennials with a well developed over-wintering structure which in the genus *Allium* is a tunicate bulb. Some are climbers, and a few have modified leaves typical of xerophytes, but most commonly there is produced a rosette of leaves (radical leaves) followed by a scape carrying a solitary flower or a raceme which may be umbellate.

The flowers are trimerous of the general form of three outer, three inner, perianth members; two whorls, each of three, stamens; and a superior tricarpellary ovary. The ovary is trilocular with axile placentation. Flower size varies considerably, 2 mm–10 cm, and there is much variation in the colour of the tepals. In a few there is a tendency towards zygomorphy. The fruit may be a capsule or a berry.

The bulbiferous genera are thought to be the most advanced, being derived from rhizomatous progenitors, and those with tunicate bulbs being more advanced than, for example, *Lilium* which has a bulb consisting of imbricate fleshy scales, open and not covered by a specialized coat. *Allium* possesses species in which there has been substantial development of the bulb and these have provided us with the onions and garlic while other species not possessed of bulbs are grown for their leaves or leaf bases which are garlic or onion flavoured. A number of *Alliums* have showy flowers and are grown as ornamentals.

Onions have been grown as a food crop for some thousands of years but only seldom have they been employed as a staple in the diet. Very high yields of bulbs, in excess of 40 tonnes per hectare, are possible and modern varieties can be grown at higher latitudes and in wetter climates than the traditional types with the result that mild, i.e. less pungent, cultivars might be grown for purposes other than adding flavour to dishes. The development of these newer, physiologically distinct, cultivars has meant the possibility that Great Britain, at one time importing about 225 000 tons per annum, will become self-sufficient in onions.

The genus *Allium* contains about 300 species and is centred in Russian Asia. There is various development of bulbs, from hardly discernible to prominent and multiplying vigorously. When present the bulb is rich in sugar. All parts of the plant have a characteristic odour and are lachrymatory. The leaves, arranged distichously, are radial, with a sheathing base, which may or may not swell. The terminal region of the leaf may be a flattened blade with parallel venation or it may be radial appearing as a hollow, tapering cylinder, somewhat flattened on the inner side. At flowering a robust scape is produced. This scape may be solid or hollow, is usually circular in section and has a mid-region which is inflated. The scape terminates in an umbellate inflorescence subtended by membranous united bracts (usually two) which act like a spathe enclosing the young developing flower head. The bracts are split as the umbel enlarges.

The individual flowers are small, 5–10 mm across and have papery tepals coloured purply-red or yellow or are almost colourless, but in the last case with perhaps a tinge of green. The flowers have the structure described for the family. The fruit, containing two seeds per ovule, is a loculicidal capsule.

The axis proper is a small conical structure giving rise to storage and functional leaves. In some species a rhizome develops to the detriment of bulbs. The production of offsets is commonplace. Sometimes bulbils are produced in the inflorescence.

Propagation of *Alliums* is by seeds or bulbs or by a combination of both (Fig. 49). Seeds are oily with a small thick-walled endosperm and are the most pungent part of the plant. They lose viability quickly and should be stored in cool, dry conditions. Germination in the field is protracted, which can lead to serious weed problems in the crop. The cotyledon is characteristically kneed with its tip remaining enclosed within the testa while it elongates to come above ground. The lower sheathing part of the cotyledon grows more rapidly than

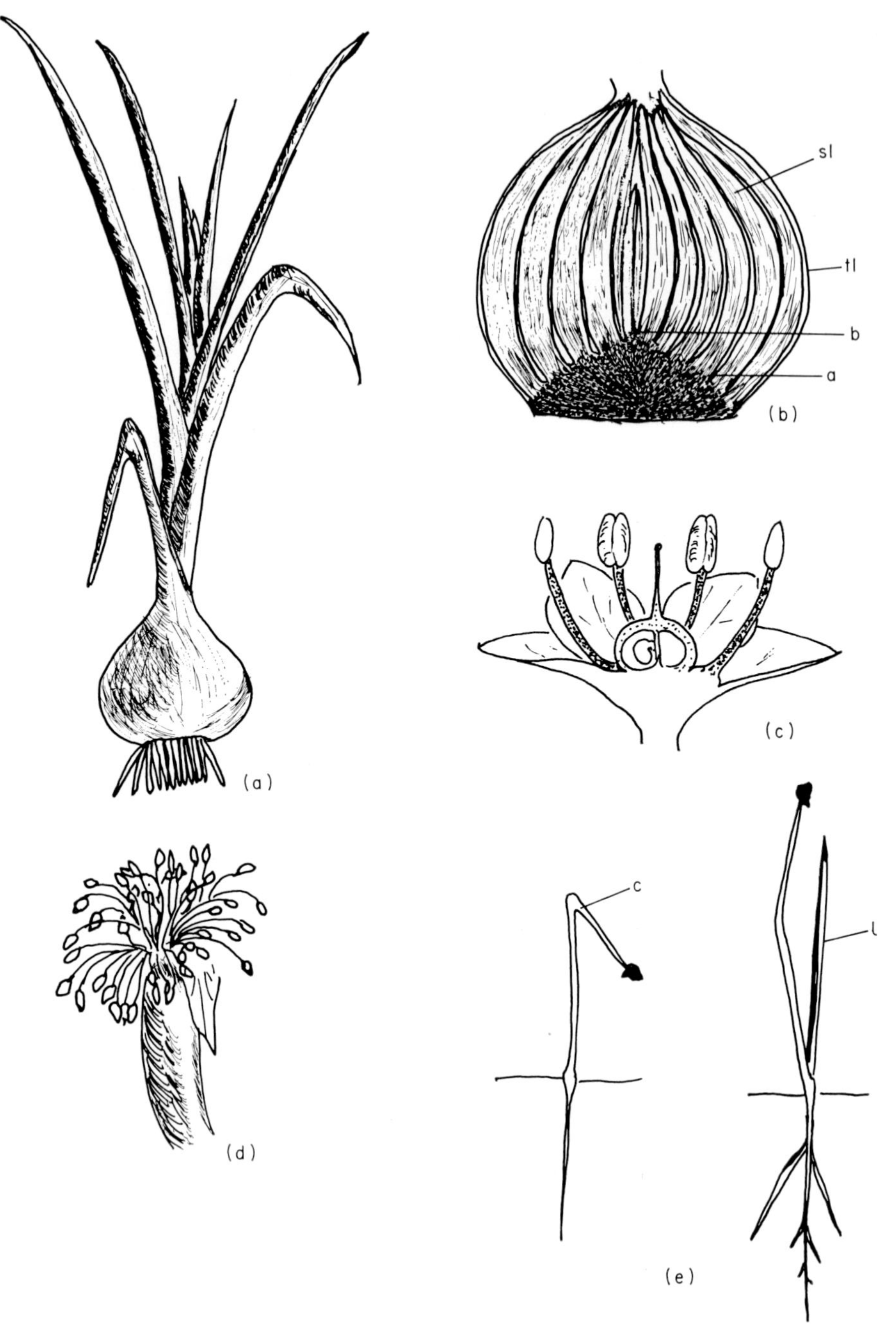

Fig. 49. Onion. (a) Adult plant. (b) Section of bulb: a, axis; b, bud with vegetative leaves; s.l., storage leaves; t.l., tunicate leaves. (c) Half-flower. (d) Inflorescence. (e) Seedlings: c, cotyledon; l, first leaf.

the upper region so bringing the seed above ground. When cotyledon extension ceases the knee straightens and the first true leaf bursts through the cotyledonary sheath. The radicle grows downwards anchoring the seedling in the soil. Soon adventitious roots develop from the short axis and this root system sustains the plant. Once established, growth is rapid. Many species are photoperiodically sensitive, but some respond not to daylength but to temperature with regard to flower induction. At high temperatures bulb production is inhibited.

The following key to the economically important bulbiferous members of the genus is modified after Purseglove.

	Leaves radial	B
	Leavcs flattened	BB
B	Leaves robust, more than 1 cm in diameter	C
	Leaves slender, less than 1 cm in diameter	CC
C	Bulb prominently flattened without tapering neck	*Allium cepa* (onion)
	Bulb not so, plant with tapering neck	*A. fistulosum* (Welsh onion)
CC	Bulbs not well developed, scape hollow	*A. schoenoprasum* (chives)
	Bulb present, scape solid	*A. chinense* (rakkyo)
BB	Bulbs an aggregate of single independent swollen storage leaves	*A. sativum* (garlic)
	Bulb, if present, consisting of two storage leaves, small axillary bulbs sometimes present	*A. ampeloprasum* (leek)

Onion belongs to *A. cepa* L., but within this species there is much variation. This species is not definitely known as a wild plant and the variants have obviously been selected by man. It would be expected that any bulbiferous perennial would in the course of years produce a clump of bulbs. Onion is grown from seed as either a winter or spring annual which means that each plant produces but a single bulb. The cultivars of this habit have been allocated to the var. *cepa*. Other cultivated forms do produce clumps of bulbs and are propagated by bulbs. These are given separate varietal status as var. *aggregatum* G. Don. To this variety belong the potato onion, ever-ready onion, and shallot (Fig. 50). That all these should be conspecific with onion is justified by the ease of hybridization between them. A third variety, *proliferum* Targioni-Tozzetti, in which inflorescence bulbils are produced is recognized. Because *proliferum* seldom produces viable pollen, hybridization studies are difficult to undertake.

All of the onions are daylength-dependent for the production of bulbs, and behave in an analagous fashion to long-day plants. With regard to flowering, the plants require to be exposed to the requisite low temperature at the appropriate stage of development. If the genotype responds to low temperature early in its development bolting is frequent but if late then it is possible to grow the

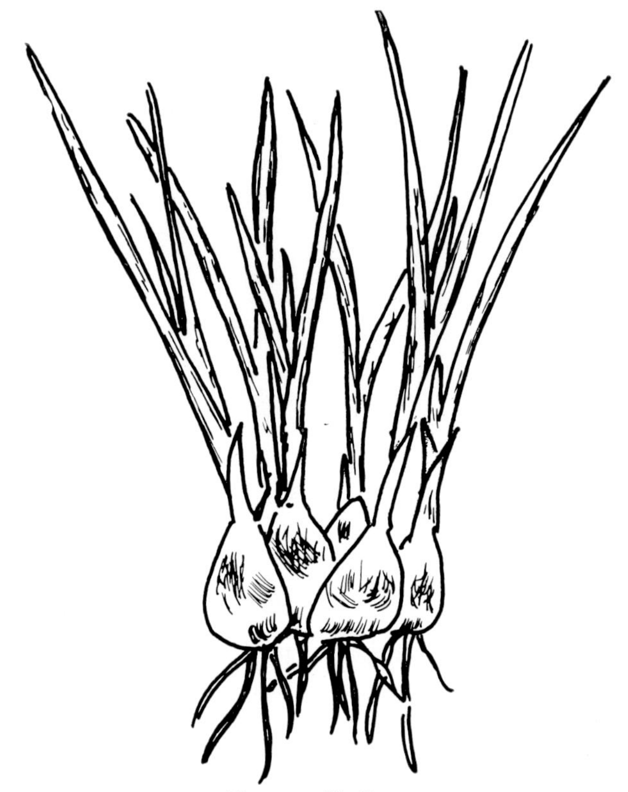

Fig. 50. Shallot.

plants over winter and so treat them as winter annuals. It can be seen that the physiological basis for the pattern of growth is quite different from the cereal, but the husbandry adopted can be the same. Low-temperature-induced flowering can be reversed by exposing the plant to high temperatures. This allows the use of 'setts' in the propagation of onions.

Normally onion is propagated by seed but if the plants are brought on in crowded conditions small bulbs are formed. These can be lifted, dried, stored, then planted out. These small bulbs are 'setts'. It is possible for the sett to be induced to flower, and if planted an increase in the size of the bulb does not take place, a scape forming almost immediately. By treating the sett with hot water the low-temperature induction is reversed. Some genotypes, however, do not require this because at the time of sett harvest they have not reached a developmental stage that is sensitive to low temperature.

In all cases, the longer the growing period prior to bulbing the greater is the crop potential. Bulbing is the result of swelling of storage leaves at the basal zone of the axis. Any factor which results in low net levels of photosynthate will reduce bulbing. High temperatures, low levels of sunshine, excess nitrogenous fertilizers, or a small area of photosynthesizing tissue, can individually or collectively contribute to the problem of non-bulbing. For successful onion culture it is necessary to grow the genotype that can best accommodate the environment of the district.

Of the flat-leaved *Alliums* leek is grown in the cooler, wetter regions, and garlic in warm dry climates. Leek, *A. ampeloprasum* L. var. *porrum* (L.) Gay, is grown as a biennial, since it is winter hardy, and grows continuously, bulbing even at low temperatures. Normally the fleshy base of the leaf does not swell to give a bulb, though sometimes in the region of the stem proper the basal region is of larger diameter than the enfolded leaf bases higher up. Late in the season axillary buds sometimes develop to give offsets.

Garlic, *A. sativum* L. produces a cluster of storage leaves from the 'clove' which is planted. The 'clove' in fact consists of a flattened axis with a single thickened leaf and a bud. The whole is protected by a papery membranous leaf. On planting, adventitious roots arise from the base of the axis, the bud grows out and an aggregate of new 'cloves' is produced by development of the basal axillary buds. This aggregate is the garlic bulb. Garlic may or may not flower.

The value of onions etc. as crops depends on how easily the particular type can be harvested and stored and also on the level of acceptance in the community. Before the advent of artificial drying and forced air ventilation, onions were grown in regions which were climatically 'safe' at harvest. Such regions had an assured dry period at the time the onion plants' leaves were drying off. Leeks, on the other hand, could be stored *in situ*, being lifted as required for the fresh vegetable market.

The odour of onions on the person is more or less acceptable, according to culture. All *Alliums* are considered to have some medicinal value, but if a person consumes large quantities of the more pungent types the onion odour, caused by propyl disulphide and other disulphides, is manifest by excretion through moist skin, especially around the mouth. The adage says it most succinctly: 'Onions—they build you up physically but let you down socially'.

CHAPTER 15

The Rosaceae: Apples and Pears

Most of the temperate fruits belong to this family. It is listed as having about 100 genera and 3000 species distributed mainly in the temperate region of the Northern Hemisphere. Four genera, *Pyrus*, *Rubus*, *Prunus* and *Fragaria* have provided us with the top fruits: pear and apple, the soft fruits; raspberry and blackberry, the stone fruits; plum, cherry, nectarine, peach and almond; and the soft fruit strawberry respectively. Many other genera have been brought into cultivation because of their showy flowers and the genus *Rosa* has provided us with one of the very best of the flowering shrubs.

The members of this family are usually perennial with many of them shrubs or trees. Their leaves may be simple or compound and are carried alternately on the stem. Stipules are present but they may be fused to the petiole. Flowers are carried in terminal racemes or in cymose inflorescences. A great variety of form is encountered in the flowers.

While the flower is basically pentamerous this is not immediately obvious in the androecium. There is multiplication of the whorls so that in some species they may be as many as four rows of stamens. Within a whorl division of the primordia can give rise to ten stamens so that it is possible for a flower to have as many as 40 stamens. It is incorrect to indicate that there is an indefinite number of stamens in this family since in those species in which there are many stamens in the flower the number present is regular. The flower is distinctive in other ways. The calyx is provided with an epicalyx lying between the main sepals, and because the epicalyx members join the sepals there results what is essentially a calyx tube. The five free petals alternate with the five sepals and then the androecium of various forms is followed by a gynoecium of from one to an indefinite number of carpels variously arranged. The way in which the gynoecium develops is dependent on the form taken by the receptacle. All species are to some degree perigynous, with the receptacle flattening to a disc, on the margins of which are carried the sepals, petals and stamens. The carpels are carried towards the centre of the disc, or even on a tumescent outgrowth which arises in the central part of the receptacle. The disc may be insignificant, or it may be enlarged and turned upwards to give an urn-shaped structure. In a few species this upturned receptacle fuses with the carpels, and there is produced a situation in which the carpels are considered to be inferior, a condition seen in apple. The fruits produced are either achenes or drupes, or are aggregate fruits of either of these, or even a false fruit in which the receptacle swells to a succulent mass (Fig. 51). In apple and pear the fruit consists of the carpels embedded in swollen receptacle and this special case is called a *pome*.

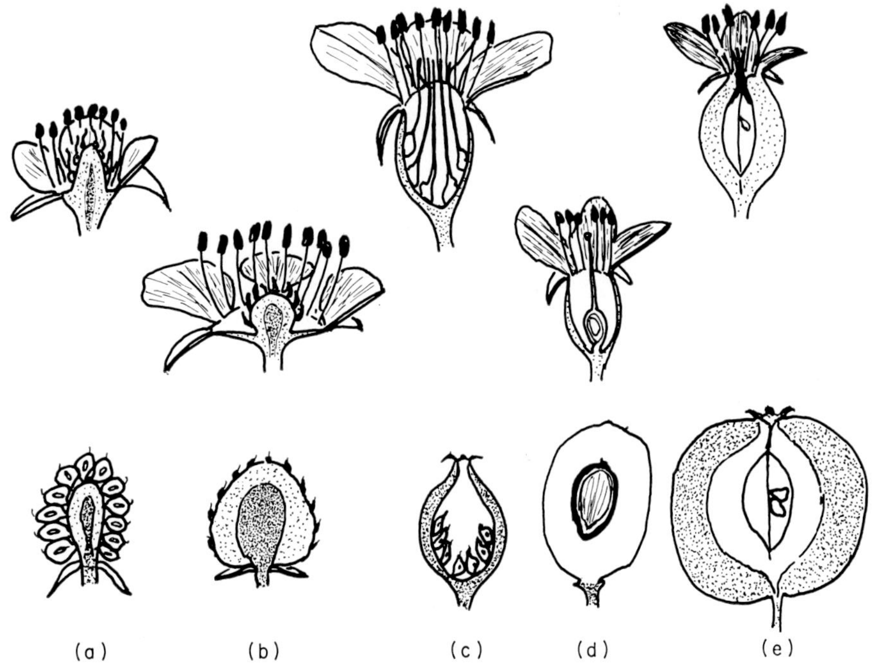

Fig. 51. Upper row, half-flowers; lower row half-fruits, of (a) *Rubus idaeus* (b) *Fragaria ananassa* (c) *Rosa canina* (d) *Prunus domesticus* and (e) *Pyrus malus*. Receptacle, light stipple; axis, heavy stipple.

The family is divided into the following subfamilies which are erected primarily on the form of the gynoecium.

Spiraeoideae: carpels arranged spirally free up to 12 but most often five; fruit dehiscent.

Rosoideae: carpels usually indefinite free enclosed within the upturned receptacle; sometimes only 1 carpel which is stalked; fruits indehiscent.

Maloideae: carpels 5–2, sometimes 1, and united to inner wall of upturned receptacle; fruit fleshy and involving the receptacle in its formation.

Prunoideae: single carpel; fruit at maturity free of receptacle and fleshy.

Many modifications of habit which aid perennation are found in this family and most of the important ones are illustrated by the crop species.

The tribe Potentilleae of the subfamily Rosoideae has provided man with two of the most important of the soft fruit genera, viz., *Fragaria* and *Rubus*.

The Genus *Fragaria*

The cultivated strawberry *Fragaria ananassa* Duchesne arose by hybridization between *F. chiloensis* Duchesne and *F. virginiana* Duchesne. This hybridization probably took place in Europe after these two New World species were brought together during the period of plant importations of the 18th century. The common woodland strawberry, *F. vesca* L., is widespread in Europe but its production of fruit is small and unpredictable. The new hybrid with a flavour like pineapple (hence the specific name) and indeed originally called the Pine strawberry, regularly bore more fruits than *F. vesca* and could be grown as a crop, albeit at first as a luxury. One of the modern varieties is claimed to yield 100 tonnes per hectare in California, and yields of 50 tonnes per hectare are commonplace in Europe. Amongst modern varieties many are able to be preserved by canning or freezing, and others have been bred for jam making. Strawberry is adaptable, relatively easily harvested and, with the prospect of suitable harvest machinery, likely to be produced more cheaply. This makes it the most widely grown of the soft fruits with the acreage under cultivation increasing. This contrasts with raspberry.

The strawberry grows as a perennial, propagating by runners. In essence it produces two types of axes, a short shoot system which branches to develop a 'crown' and a long shoot system which is in fact the runner. Inflorescences are carried on the short shoots. This distinction in the morphologies of the fertile and sterile axes is manifest in all of the Rosaceous fruits but in the tree species there is not the same obvious distinctly different function of them. The type of axis produced and the capacity to flower is strongly influenced by the environment.

Seedlings arise in the wild, but in commerce young plants obtained from runners are planted. These plants from runners are called first year plants or 'maidens' and consist of the terminal region of the runner which has short internodes and a well-developed terminal bud. This terminal region of the runner is subtended by two long internodes. The node separating these internodes carries a scale leaf and the bud at the axil may or may not grow out as a long shoot or remain short. Adventitious roots can develop at this node. The runner is plagiotropic though at first, when produced from the mother plant, it exhibits some degree of negative geotropism. After the second internode on the runner subsequent internodes are short and at each node a stipulate, petiolate, ternate leaf is produced. The petioles at this stage are erect, upright and relatively long. The leaflets are coarsely toothed and relatively small. On the underside the leaflets have fairly long hairs but are glabrous blue-green above. Adventitious roots are formed and this terminal region of the runner becomes independent of the mother plant. This whole process of establishment of the runner takes place after flowering during late summer. A maiden with the long upright petioles and relatively small leaflets exhibits the vegetative growth form. Flower induction occurs in the maiden during summer in response to long days.

On overwintering, the leaves of the vegetative growth form die back, and on the resumption of growth the leaves that are produced have short petioles, are not upright, with leaflets that are relatively large and dark green. The flower buds open and the plant exhibits the flowering growth form. It has been found that the transition from one growth form to the other will only take place if the plant is exposed to low temperatures after flower induction. If it is kept at high temperatures only vegetative growth takes place.

The flowers are carried on cymes, at first dichasial, later becoming monochasial in the secondary or tertiary branches. Since the primary flower produces the largest fruit and fruit size decreases with advancing order of the inflorescence, it is undesirable for a single plant to have a small number of large inflorescences. In terms of producing an evenly ripening crop of standard sized fruits it is better that the plant produces a larger number of smaller inflorescences that develop synchronously. The latter arrangement aids mechanical harvesting.

The individual flowers have the sepals, petals and stamens arranged on the rim of a flattened torus. Central to these in the middle of the torus carpels are carried on a spherical mound. Basically the flower is pentamerous but in some cultivars there are more than five sepals and petals. There is an indefinite number of carpels. The true fruit is an achene derived from each of the single-ovuled carpels but these achenes are embedded in a succulent tissue which comes from the torus mound. The expansion of the torus is conditional upon fertilization, for growth substances produced during embryo development incite expansion of the torus cells. The production of this false fruit, i.e. the succulent torus mass, is restricted to the immediate vicinity of fertilized ovules, and if these are not distributed evenly over the torus misshapen 'fruits' result. Pollination is important and, when the crop is grown under protection, it is necessary to provide pollinators. In the U.K. blowflies have been found to be satisfactory pollinators for 'tunnel' grown plants.

In the United Kingdom maidens are deflowered and the plants are cropped for two years subsequent to establishment. After this period of three years the crop is replaced. Replanting at what seems to be frequent intervals is found necessary to reduce the incidence of soil-borne virus infection.

Some cultivars will flower in autumn to give a second crop. This happens when the plant is early and flower initials are laid down in summer. The so-called ever-bearing strawberries are not photoperiodically sensitive, so can produce flowers whenever the temperature is high enough and there is sufficient light for active growth. Recent work has shown that while runners are forming and shoots are growing vegetatively an inhibitor of flower production is produced by the leaves. It is thought this inhibitor is a gibberellin. Growers have long known that removing leaves of plants which had fruited increased the following year's crop. The physiological bases of this practice has been found with the discovery of the inhibitor in vegetative plants.

The Genus *Rubus*

Many members of the genus *Rubus* are cultivated for their fruits. These plants are generally spiny shrubs with characteristic separate vegetative and flowering axes, but within the genus there are species with annual stems either herbaceous or woody. The latter are often abundant locally and the fruits of such wild plants are gathered (e.g. *R. chaemaemorus* L., cloudberry, and *R. saxatilis* L., stoneberry or stone bramble).

The most widely grown species is *R. ideaus* L., the raspberry. Raspberries are perennials with biennial aerial shoots. The shoots arise adventitiously from a robust perennial root system, and thus they might be likened to suckers. Large clumps form if plants are unattended, but in culture each plant is pruned to a set number of shoots, the actual number depending on the method of culture. The sucker in its first year produces an upright cane carrying alternate pinnately compound stipulate leaves, the leaflets being irregularly toothed. The terminal leaflet is larger than the laterals, and all are generally ovate to ovate lanceolate and are densely hairy below. The base of the sucker is clothed in stiff prickles which are lost as the season advances. Along the length of the shoot there are more robust thorns varying in density according to the cultivar.

During the first year of growth there is no obvious indication of inflorescences, but in fact the axillary buds, at the onset of dormancy, contain inflorescence primordia. The date at which primordia are laid down depends on the cultivar and the environmental conditions, and in some cultivars the primordia are laid down early, growing on during the season, and the cane in its first year carries flowers. These are the autumn-fruiting raspberries. The inflorescence primordia overwinter and in the second year of the cane's growth the axillary buds grow to give short shoots terminating in a cymose inflorescence carrying up to 10 flowers.

In *Rubus* there is no epicalyx, and the five sepals are ovate lanceolate with long acuminate tips. The petals, equalling in number and as long as the sepals, are oblong, white and erect. There are many stamens, and numerous spirally arranged carpels on a swollen torus. Each ovary has two ovules. The fruit is a drupel and adheres to the receptacle to result in an aggregate fruit. Fruits, which are pubescent, are most commonly red but may be yellow or white.

The wild blackberry (bramble) closely resembles raspberry, but is more robust, and does not produce inflorescence primordia until late in the year, in some even after leaf fall. This results in a plant which flowers and fruits later in the season. Blackberry, *Rubus fruticosus* L. agg. differs from *R. ideaus* by having palmately compound leaves, and in the canes tending to be more persistent. It is a variable species due to a peculiar breeding system and hybridization, and because of the constancy of features of clones many of the variants have been given specific rank. The British *Rubi* are listed as having 391 species by Watson but Clapham *et al.* in their treatment of the British Flora list over 700 species. Doubtless there are many clones that could be described as species but there is limited value in adopting this taxonomic approach.

Hybrids between *idaeus* and *fruticosus* can be formed as can hybrids between *R. ideaus* and other *Rubi*. Loganberry is such a hybrid, between *R. ideaus* and *R. ursinus*, Chan. et Schlecht var. *vitifolius*.

The Tree Fruits

The stone fruits all belong to the genus *Prunus*. This genus of some 200 spp. is distributed mainly throughout the North Temperate zone. A few occur in the Andes. They are small trees or shrubs sometimes armoured eventually producing a more or less succulent fruit. Five species are widely grown for their fruits:

Prunus armeniaca L., the apricot
P. cerasus L., sour cherry (the cherry of commerce)
P. communis Batsch., the almond
P. domestica L. ssp. *domestica*, plum
P. persica (L.) Batsch., peach and nectarine

Others are important locally and many are cultivated for their showy flowers.

The general description of *Prunus* which follows is characteristic not only of the stone fruits but also of the top fruits (the genus *Pyrus*). In both these classes of fruit the general management is governed by the morphology and physiology of the tree.

Seed germination in Rosaceous trees is not uncommon, nor is it particularly difficult to accomplish in practice. When shed the seed is dormant, and in many cases this is due to a physiological immaturity of the embryo. After-ripening of this immature embryo takes place at low temperatures, such as those prevailing in the soil over winter, and these same conditions will effect the breaking of dormancy. If the seed is induced to germinate or germinates before the completion of after-ripening, a 'dwarf' seedling results (Fig. 52c). This seedling has short internodes and small leaves. Gibberellic acid, if applied either to the seed or the dwarf seedling, will produce a seedling of normal stature in the absence of after-ripening.

The seedlings which result are dissimilar and for this reason most Rosaceous fruits are propagated by cuttings, or as is most frequent by grafting proven scions on to rootstocks. At one time the rootstocks used were seedling rootstocks but as a result of the pioneer work carried out by the East Malling Research Station careful observation of the scion–rootstock interaction resulted in grafts being made between proven scions and proven rootstocks. The rootstocks are propagated vegetatively by layering or from suckers.

Fig. 52. (a) Tip-bearing character in apple. (b) Spur bearer in apple. (c) Dwarf and normal pear seedlings.

The scion–rootstock interaction is an interesting phenomenon, best understood by considering the extremes of reaction that have been observed. At one extreme there are the dwarfing rootstocks which when used give rise to a plant which is small, seldom exceeding 4 m, produces fruit buds early in its life, often after only two years' growth, and is of short duration, exhibiting senescence after perhaps 20 years. The 'vigorous' rootstocks give a plant which is robust, reaching 8–10 m, produces fruit buds only after an extensive period (15–20 years), and is long-lived. In establishing orchards and plantations, account must be taken of the rate of return on capital when deciding what type of plant to establish. The other economic factor to consider is the cost of labour, since harvesting the dwarf types requires fewer man hours than is the case with robust large trees.

Modern systems of plantation management and arrangement give higher and earlier yields when the scions are grafted on dwarfing rootstocks.

Before the widespread use of the dwarfing rootstock, vigorous, non-productive trees were root pruned and/or girdled. These practices depauperated the tree, and after such a treatment it was often found that fruit buds were formed instead of vegetative buds. It is now known that an application of gibberellin to some Rosaceous trees promotes the formation of vegetative buds, whilst treatment with compounds which prevent the plant from synthesizing gibberellin promotes the production of flower buds. It is reasonable to conclude that the grafting on to a particular rootstock will influence the hormone economy of the scion towards conditions conducive to either fruit or vegetative bud production.

The small saplings, either from seed or grafts, produce a main stem and while the tendency is for this to develop to give a well-developed bole the stage at which branching takes place is genetically determined. In culture, pruning is employed to 'shape' the tree and many special pruning procedures have been developed to give rise to forms such as cordons, espalier, fans, spindle-bush, etc., as well as the more often encountered bush, half-standard and standard shapes. When pruning, the morphology of the plant being pruned must be taken into consideration.

Two distinct types of shoots are found in Rosaceous fruit trees: long shoots terminating in a vegetative bud and often free of axillary branches (but when present the axillary buds are vegetative), and short shoots terminating in a flowering bud. Vegetative buds are acutely pointed and narrow at the base, whereas the flowering bud can be almost globular, but is always obtuse with a broad base. Long shoots have extensive internodes; the short or spur shoots have internodes so small as to be hardly measurable.

Short shoots may terminate a branch or they may occur as small side branches. Plants of the former morphology are tip bearers, and plants with lateral short shoots are called spur bearers. Pruning regimes seldom involve the removal of short shoots, which means that the pattern of pruning differs for these two morphological forms.

The Genus *Prunus*

Leaves on the long shoots are borne spirally and each leaf is petiolate, stipulate but stipules falling early, and with an ovate, toothed blade. Members of this genus have extra-floral nectaries at the junction of the base of the blade and the petiole, and sometimes in the angle between the main and subsidiary veins. On the short shoots the leaves are carried on tight spirals and are generally smaller, with shorter petioles, than those of the long shoots. If the short shoot is floriferous the number and size of the leaves are reduced still further so that there may be as few as two small leaves associated with the inflorescence. Flowers are carried in racemes or umbels, the latter being the case with the cultivated species.

The individual flowers have five sepals, five petals, 15–20 stamens all arranged on the edge of a cup-shaped receptacle. There is a single carpel on a short stalk rising from the base of the cup. Only one of the two pendulous ovules matures to a seed, and the pericarp becomes large and fleshy to give a drupe. The cup is insignificant when the fruit is mature. The drupes may be shiny and covered with a bloom (wax plates), or pubescent and vary in diameter from about 1.5 cm for small cherries to 8 cm for large peaches.

The species are variously self-fertile or self-sterile.

It is always essential to have the appropriate pollinator for the main cultivar which is grown. In the case of self-fertile cultivars, fertilization does not present any genetical problems, but with self-sterile cultivars a compatible genotype must be grown to flower at the same time as the main crop.

If compatible pollen is available at flowering, fruit set will depend on climatic and biological factors prevailing at that time and immediately after pollination. It is necessary to have an adequate population of pollinating insects and it is recommended to introduce beehives into orchards at flowering time. Viable pollen and healthy stigmas and styles are essential and under normal climatic conditions there are few problems in this area, but low temperatures and desiccating winds can result in non-viable pollen and also shrivelling of the style both before and after pollination but before the pollen tube reaches the embryo-sac.

Too high a fruit set results in too many small fruits, and it is sometimes necessary to thin the fruits to obtain premium quality produce. With smaller fruited types, e.g. damsons and cherries this practice of fruit thinning cannot be justified economically.

If a tree fruits vigorously it is often found that the production of flowers, as observed in the following year, is reduced and so will be the crop. This can institute a tendency towards biennial bearing, the tree carrying alternate heavy and light crops. This tendency is understood when it is realized that these stone fruits are long-day plants being induced to flower by summer daylengths and then initiating flowers at the same time as fruits are ripening, these fruits being derived from flowers laid down the previous year. There is competition for nutrients between flower initials and expanding fruits, and if there are many of the latter the former are deprived of food materials and fewer are produced. A contrary situation will exist the following year—few fruits, many flower initials.

Even production can be achieved by the grower if he manages the crop by fruit thinning or by ensuring that his trees do not behave synchronously.

The genus *Prunus* is divided according to the manner of folding of the leaves in the bud,the overall form of the calyx tube, and the nature of the inflorescence. The main cultivated species are separated as follows:

P. armeniaca

Leaves rolled in bud; tree to 6 m with smooth branches; leaves ovate, pointed, markedly toothed, without glands but with down in angle between midrib and lateral veins, to 10 cm; petioles with fine hairs; flowers large, 2.5 cm, white in umbels of two to five. Fruits to 2.5 cm or larger, pericarp yellow turning red, mesocarp yellow.

P. cerasus

Leaves folded in bud; tree to 6 m, rounded in shape, branches smooth; leaves elliptical, smooth, finely dentate, prominent extra-floral nectaries, to 7.5 cm;

flowers to 2.5 cm in clusters, reducing in number; fruits to 1.5 cm, sometimes larger, pericarp dark red to purple, flesh various, yellow to claret, soft and acid; 'stone' round or oval; a species with many named varieties grown both for fruits and as ornamentals.

P. communis

Leaves folded in bud; small tree between 3 and 7 m, at first erect but later becoming rounded; twigs smooth; leaves large, elliptical 7.5–15 cm finely toothed, extra-floral nectaries prominent; petiole short 2.5 cm. Flowers to 5 cm solitary almost without peduncle. Fruit elliptical, flattened, pubescent. Mesocarp dry splitting to expose the stone which is smooth with distinct elongate pits.

The variety *amara* is the bitter almond and *dulcis* is the sweet almond.

P. domestica

Leaves rolled in bud; small bushy tree with pubescent branches which soon become glabrous, occasionally thorny; leaves obovate, elliptical, serrate, with slight pubescence but becoming glabrous and dull above, to 10 cm. Flowers in groups (1–3), pedicels 0.5–2 cm, up to 1.5 cm across; fruit round, elliptical, occasionally pointed; pericarp yellow, red, purple, or blue black, mesocarp succulent greenish or claret; stone flattened, rough and slightly pitted. The ssp. *domestica* is the plum proper, whilst ssp. *italica* (Borkh.) Hegi is the greengage. Neither of these is thorny. In greengage the flesh adheres to the ovoid stone, but in plum the flesh is free from a flat, sharply-angled stone. The third ssp. is *insititia* (L.) C. K. Schmid, Bullace, which is usually thorny and has a small fruit. Damsons are the cultivated form.

P. persica

Leaves folded in bud; bushy tree with smooth branches; leaves lanceolate, slightly curved, dentate, up to 15 cm and with small petiole, 1–2 cm; flowers singly or in pairs to 3.5 cm across on small peduncle; fruit round with cleft, sometimes compressed, at maturity large to 7.5 cm in diameter and may be pubescent (peach) or smooth (nectarine); pericarp yellow, suffused with red where exposed to sun; mesocarp pulpy, sometimes stringy, yellow or pinky-yellow; stone wrinkled and pitted, either easily removed (free stone) or with mesocarp adhering (cling) at maturity.

Other species with many named varieties are grown as ornamentals.

The Genus *Pyrus*

I am including within the genus *Pyrus* both pear, *P. communis* L. and apple, *P. malus* L. The apple has been allocated to the distinct genus *Malus* as *Malus sylvestris* Mill. (=*M. pumilis* Mill) by some authors and these authors consider the genus to have some 25 species. These 25 spp. do constitute the subgenus *Malus* of *Pyrus* and if included within *Pyrus*, as here, then *Pyrus* has some 50 species.

In *Pyrus* the habit of the plant is like that described for *Prunus*, and all the features of the morphology and physiology that were stressed in the previous section apply here, viz. stock–scion interaction, response to daylength, biennial bearing, pollination and compatibility.

The fruit of *Pyrus* is a pome but there are differences between apples and pears, and the related quinces of the genus *Cydonia* (the flowering quinces are most often placed in the genus *Chaenomeles*).

In *Pyrus* there are two ovules per carpel but in *Cydonia* more than two are present in each loculus. The subgenus *Malus* (of *Pyrus*) is recognized by having the styles connate at the base and the operculum of the pome remaining open. *Chaenomeles* differs from *Cydonia* in having styles connate at their base, but in both these genera the pome is closed at the top. If the connation of the styles is a good generic character then *Malus* should be separate from *Pyrus*. The degree of genetic relationship among the members of *Pyrus* (*sensu lato* as here) and *Cydonia* and *Chaenomeles* is difficult to determine by hybridization studies, but that there is a close affinity between them is evidenced by the fact that grafts between pear and quince are possible; indeed quince is the preferred rootstock for pears, and when a direct graft between a rootstock and a scion cannot be made, a bridging graft using another genus between rootstock and scion often allows the graft union to be accomplished.

The general description of pear is: medium-sized tree to 15 m, generally erect, sometimes with thorns; young twigs glabrous or almost so, and if pubescent, softly yellowish brown; leaves with roundish blades, orbicular to ovate with subcordate base (2.5–6 cm long), serrate along the margins, tomentose when young but becoming glabrous, though sometimes retaining some fine hairs on the lower surface. The petiole is as long as the blade. The flowers are carried on a corymb which, because of the contraction of the axis, appears umbellate; flower stalks up to 3 cm; flowers up to 3 cm across, petals white, anthers purple; fruit large, tapering to base, sepals persisting at apex and with stone cells in flesh.

A general description of the apple is: small tree or shrub to 10 m and generally with rounded crown, sometimes thorny; young twigs more or less pubescent; leaves with ovate or oval blades rounded at base (3–4 cm long), serrate along the margins. The petiole is short, half as long as the blade. Flowers are carried in groups forming almost a true umbel; flower stalks 1–3 cm; flowers to 5 cm across with white petals tipped with pink along margin, anthers yellow. Fruits more or less globular with sunken apex and base, coloured green, yellow or red, sometimes particoloured. Calyx persistent, stone cells absent.

Modern methods of growing top fruits attempt to arrange the plants, and design the shape of the trees, so that there is maximum interception of light without there being substantial mutual shading. The traditional orchard with large trees each with a large crown establishes both sun and shade conditions, the former on the periphery of the crown and the latter within the crown. The shade part of the plant is essentially parasitic on the productive illuminated part. Bush forms go part of the way towards reducing the proportion of the plant in shade, but cannot be planted to give optimum interception of light. Newer orchard systems space the plants to form hedges pruned to a triangular section and with the hedges in arrays of twos running in the best compass direction for the latitude. The ultimate is the so-called meadow orchard where the trees are grown on extreme dwarfing stocks, treated with dwarfing agents and encouraged in biennial bearing. The single stem produced on the close-growing plants is cut and the fruits removed by pickers in protected conditions. A single stem is allowed to grow from each stool and the cropping pattern repeated on a two-year cycle.

These developments illustrate how a knowledge of the morphology and physiology of a plant can be used to tailor-make a crop.

Bibliography

This list is not exhaustive and for the advanced student, reference to the bibliographies of some of the specialist texts will provide entry to the very extensive literature that applies to the crop plants.

General

Tropical Crops: Dicotyledons. J. W. Purseglove, Longmans, London, 1968.

Tropical Crops: Monocotyledons. J. W. Purseglove, Longmans, London, 1972.

Syllabus der Pflanzenfamilien. A. Engler revised by H. Melchior Bd. II, Gebrüder Borntraeger, Berlin, 1964.

A Dictionary of the Flowering Plants and Ferns. J. C. Willis revised by H. K. Airy Shaw, C.U.P., London, 8th Ed. 1966 (earlier editions still useful).

Index Kewensis and Supplements. O.U.P. London, 1893 *et seq.*

Die Natürlichen Pflanzenfamilien. A. Engler and K. Prantl, Verlag von Wilhelm Engelmann, Leipzig, 1889 *et seq.*, and more recent editions.

The Families of Flowering Plants. J. Hutchinson, O.U.P., London, 3rd Ed. 1973.

The Genera of Flowering Plants Vols I and II. J. Hutchinson, O.U.P., London, 1964.

Manual of Cultivated Plants. L. H. Bailey, Macmillan, New York, 1949.

Dictionary of Gardening (Royal Horticultural Society) edited by F. J. Chittenden and P. M. Singe, O.U.P., London, 4 Vols. and Suppl. 1956 *et seq.*

Palaeoethnobotany. The Prehistoric Food Plants of the Near East and Europe. J. M. Renfrew, Methuen, London, 1973.

Agricultural Botany, Theoretical and Practical. J. Percival, Duckworth, London, 8th Ed. 1936.

Evolution of Crop Plants. N. W. Simmonds, Longmans, London, 1976.

Chapter 1

Grass Systematics. F. W. Gould, McGraw-Hill, New York, 1968.
The Gramineae. A Study of Cereal, Bamboo, and Grass. Originally C.U.P. London, 1934, reprinted Wheldon and Wesley, Codicote, 1965.
First Book of Grasses. A Chase, Smithsonian Inst. Press, Washington, D.C., 1959.
Manual of the Grasses of the United States. A. S. Hitchcock, revised by A. Chase, U.S. Government Printing Office, Washington D.C., 1950.
Grasses. C. E. Hubbard, Penguin, Harmondsworth, Middlesex. 1954.
British Grasses and their Employment in Agriculture. S. F. Armstrong, C.U.P., London, 1943.
The Wheat Plant. J. Percival, Duckworth, London, 1921.
Wheat: Botany, Cultivation and Utilization. R. F. Peterson, Leonard Hill (Books) Ltd., London, 1965.
Corn: Its Origin, Evolution and Improvement. P. C. Mangelsdorf, Harvard University Press, Cambridge, Mass., 1974.
Rice. D. H. Grist, Longmans, London, 5th Ed. 1975.
Cultivated Races of Sorghum. J. D. Snowden, The Trustees of the Bentham-Moxon Funds, London, 1936.
Sorghum. H. Dogget, Longmans, London, 1970.
The Sugar Cane. A. C. Barnes, Leonard Hill (Books) Ltd., London, 1964.
The Bamboos: A Fresh Perspective. F. A. McClure, Harvard University Press, Cambridge, Mass., 1966.

Chapter 2

Chemotaxonomy of the Leguminosae. Ed. by J. B. Harborne, D. Boulter and B. L. Turner, Academic Press, London, 1971.
Alfalfa: Botany, Cultivation, and Utilization. J. L. Bolton, Leonard Hill (Books) Ltd., London, 1962.
Soybeans: Improvement, Production and Uses. Ed. B. E. Caldwell, American Society of Agrinomy Inc., Madison, 1973.
Leguminous Forage Plants. D. H. Robinson, Ed. Arnold and Co., London, 1947.
Nitrogen Fixation in Plants. W. D. P. Stewart, The Athlone Press, University of London, London, 1966.

Chapter 3

The History and Social Influence of the Potato. R. N. Salaman, C.U.P., London, 1949.
The Potato and its Wild Relatives. D. S. Corrrell, Texas Research Foundation, Renner, Texas, 1962.
The Potatoes of Argentina, Brazil, Paraguy, and Uruguy. J. G. Hawkes and J. P. Hjerting, Clarendon Press, Oxford, 1969.
Tobacco. B. C. Akehurst, Longmans. London. 1968.

Chapter 4

The Natural History of Palms. E. J. H. Corner, Weidenfeld and Nicolson, London, 1966.
The Oil Palm. C. W. S. Hartley, Longmans, London, 1967.

Chapter 5

Bananas. N. W. Simmonds, Longmans, London, 2nd Ed., 1966.

Chapter 6

Cucurbits: Botany, Cultivation, and Utilization. T. W. Whitaker and G. N. Davis, Leonard Hill (Books) Ltd., London, 1962.

Chapter 7

The Biology and Chemistry of the Cruciferae. Ed. J. G. Vaughan, A. J. MacLeod and B. M. G. Jones, Academic Press, London, 1976.

Chapter 9

Cotton (with special reference to Africa). A. N. Prentice, Longmans, London. 1972.

Chapter 12

Coffee. F. L. Wellmann, Leonard Hill (Books) Ltd., London, 1961.
Cocoa. G. W. Lock, Longmans, London, 3rd Ed., 1975.
Tea. T. Eden, Longmans, London, 2nd Ed., 1965.

Chapter 13

The Biology and Chemistry of the Umbelliferae. Ed. V. H. Heywood, Academic Press, London, 1971.

Glossary of Terms

Abaxial: The side of a leaf farther from the axis, defined in relation to its origin; normally the lower surface.
Acaulescent: Not having an obvious stem.
Achene: A dry indehiscent single-seeded fruit produced from a single carpel. The fruit wall separate.
Acropetal: Polarized distribution from the base to the apex.
Actinomorphic: Said of a flower which exhibits many planes of radial symmetry. (=Regular).
Adaxial: The side of a leaf nearer the axis, defined in relation to its origin; normally the upper surface.
Adnation: The fusion of two dissimilar organs.
Adventitious: Said of organs arising from an unexpected position, e.g. roots on stems, buds on leaves.
Aestivation (**Estivation**): The manner of folding of flower parts in the bud.
Alae: Literally the wings. The lateral petals of the flower of legumes.
Albedo: The white portion of the fruit wall of citrus fruits.
Aleurone: A specialized storage tissue found in seeds, the cells of which contain characteristic protein and oil aggregates—the **Aleurone Grains**.
Alkaloid: Nitrogen-containing compounds which are slightly basic, i.e. alkali-like.
Allergen: Any substance, usually a protein, which can elicit an allergy in a mammal.
Anatropous: Ovules which are bent completely over such that the embryo sac is parallel to the stalk axis and the micropyle faces towards the stalk attachment.
Androdioecious: Said of a species in which individuals possess either only complete or only staminate flowers.
Andromonoecious: Said of a species in which the individuals possess both complete and staminate flowers.
Anemophily: The transferring of pollen from one flower to another by wind.

Annual: The completion of the life-cycle within a calendar year. May be a winter annual in which case the plant exists as a seedling during the winter months, or a spring annual, in which case the plant does not over-winter as a seedling.

Anther: The pollen-bearing part of the stamen consisting of pollen sacs.

Apocarpy: The condition in a flower where each of the carpels is separate.

Apogamy: The production of embryos without there being gametic fusion. The embryo arises directly from a cell of the maternal tissues.

Apomixis: Reproducing by seeds not resulting from fertilization; hence genetic recombination is not encountered.

Arrow: The large feathery paniculate heads of certain tropical grasses, especially sugar cane and its allies. Sometimes also used in the case of maize.

Ascending: With stems tending to an upright vertical position.

Auricles: Small distinct hardened pieces of tissue seen at the junction of the blade and sheath of certain grass leaves.

Awn: Bristle-like appendages of the lemma, and sometimes the glume, in some grasses. The awn may be terminal in which case it is a direct continuation of the lemma tip or it may be attached separately along the back near the tip (sub-apical) near the middle (dorsal), or near the base (basal). The awn should be greater than one-third of the length of the subtending structure otherwise it is an awn point. Awns may exhibit movement occasioned by changes in atmospheric moisture contents.

Axillary: Said of a structure carried within the axil.

Bacteroid: Mis-shapen bacterial cells of *Rhizobium* as found within leguminous root nodules.

Basipetal: Polarized distribution from the apex to the base.

Beak: The terminal sterile portion of any dry fruit, but particularly those of the Cruciferae.

Bearded: Any cereal head of which the floral parts possess long prominent awns.

Beardless: Cereal heads which are awnless.

Berry: A fleshy fruit in which all parts of the pericarp are succulent.

Biennial: A plant which completes its life cycle in two calender years, and which during the first is not reproductive.

Bifid: Split into two at the tip.

Bifurcated: Divided into two branches.

Biosphere: The peripheral layers of the Earth which can support life.

Bivalent: The configuration adopted when two completely homologous chromosomes pair at meiosis.

Bolls: The ripe fruit of cotton.

Bolter: (i) An individual plant of a biennial species which flowers in its first year of growth. Normally confined to root crops. (ii) An aberrant form of potato.

Bract: The leafy structure subtending a flower or flowering axis.

Bracteole: A leafy structure subtending a flower but of a lower order than a bract.
Bulb: A storage organ consisting of swollen modified leaf bases or scale leaves arising on a truncated stem. The following year's flower bud primordia are contained centrally, and in the case of tunicate bulbs the whole is surrounded by specialized outer membranous scale leaves.
Bulliform cells: Literally blister-like cells; see Hinge cells.
Bush bean: Said of any bean species or cultivar with a sympodial growth form. Contrast with pole bean.
Caducous: Tending to drop off early, and therefore not persistent.
Calyx: The outermost row of protective members in a heterochlamydous flower, and those most like leaves.
Campylotropous: Said of an ovule which is bent in such a way that the embryo-sac is curved and lies at right angles to the stalk. The micropyle and chalaza are obvious.
Capitulum: A racemose inflorescence in which the flowers arise on an abbreviated axis, which may be globose, so conferring on it a compact spherical form.
Capsule: A dry dehiscent fruit which may be produced from one or more carpels. Commonly splits either along the dorsal sutures (loculicidal) or along the ventral sutures (septicidal).
Carpel: That floral structure which contains the ovules. Imprecisely referred to as the female part of the flower.
Carpellate: Said of a flower which possesses carpels but no stamens ≡ female flower.
Carina: The anterior structure formed by the adhesion of the two anterior petals in some leguminous flowers.
Carpophore: That terminal part of an axis which carried adnate with it the carpels (see Umbelliferae).
Caruncle: A fleshy protuberance surrounding the hilum of some seeds.
Caryopsis: The fruit of most members of the Gramineae. Characterized by being single-seeded and with the pericarp fused to the testa.
Caulescent: Possessing an obvious stem.
Cauliflorous: Carrying flowers directly from the stem, not subtended by a leaf not axillary, and not terminal.
Cereal: Any member of the Gramineae the grains of which are used mainly as food for man after being milled and ground.
Chemotaxonomy: The employing of chemical constituents of organisms in determining systematic relationships.
Chitting: The practice of permitting active growth of dormant or quiescent axes to develop before planting. Used in potatoes and certain seeds, e.g. cucumber.
Chupon: The characteristic upright shoot produced by certain tropical understorey trees (see Cocoa).

Circumnutation: The phenomenon exhibited by growing shoots which results in the apex describing an opening spiral path in space.
Class: A taxon above the level of order but below that of Phylum, and comprising a large number of orders.
Cleistogamy: The condition whereby pollination is effected within the buds of (usually small) flowers.
Cline: A graded series of variants which may be correlated with other features associated with the species, e.g. ecological factors which exhibit gradual change. Usually prefixed to indicate the independent variable; ecocline in the above case.
Clone: A collection of individuals all derived by vegetative reproduction from a common ancestor and therefore each of identical genetical composition.
Coarse grain: Those grass grains grown primarily for consumption by animals. Many are now also eaten by man.
Cobs: The infructescence of maize; also used in Britain to refer to the nut of *Corylus avellana* L.—hazel.
Coir: The fibrous mesocarp of Coconut used in the manufacture of coarse matting and ropes.
Coleoptile: The prophyll of the plumular bud. A colourless hollow cylinder closed at the tip and surrounding the plumule in a grass embryo; capable of extensive growth.
Coleorhiza: A colourless sheathing tunic surrounding the radicle and seminal roots of grasses; not capable of extensive growth.
Connate: Said of like structures fused together, e.g. connate anthers as in Solanaceae.
Copra: The dried oily endosperm of Coconut.
Corm: A swollen tunicated, subterranean stem capable of perennation. Subsequent corms are produced from the upper axillary buds of the existing structure.
Corolla: The inner part of the protective perianth of a heterochlamydous flower usually delicate and coloured other than green.
Coronal root: A root developing in the crown region of a plant, particularly the adventitious roots on the subterranean portion of the grass stem.
Corymb: A racemose inflorescence such that the individual peduncles elongate to differing degrees and bring the flowers to lie in a single horizontal plane.
Cosmopolitan: Occurring in each of the recognized plant geographic regions; throughout the whole world.
Critical daylength: The daylength below which, in long-day plants, flowering does not occur. In short-day plants the daylength above which flowering does not take place.
Cross-inoculation: The process of attempting to establish a successful combination between the bacteria isolated from one group of host plants and the members of another group of host plants. Cross-inoculation groups are established when the limits of successful inoculation pairs are defined.

Cross-pollination: The transferring of pollen from the anthers of one flower to the stigmatic surface of the carpels of another different flower.
Crown: That portion of the plant's axis at and just below soil level.
Culm: The stem of a grass.
Cultivar: The name given to a commercially distinct group of individuals. Considered as a taxon in its own right.
Cyanogenetic: Possessing the capacity to produce or release hydrocyanic (prussic) acid.
Cyme: A determinate inflorescence in which the oldest flower is at the apex. Younger flowers arise on lower-order peduncles. (Adjective **Cymose**.)
Dehisce: To split.
Deltoid: Shaped like the Greek capital letter delta (Δ)—triangular.
Diadelphous: Having two sets of stamens united accordingly.
Dichasium: A cymose inflorescence which is symmetrical in its development on each side of the oldest flower. Shaped thus—

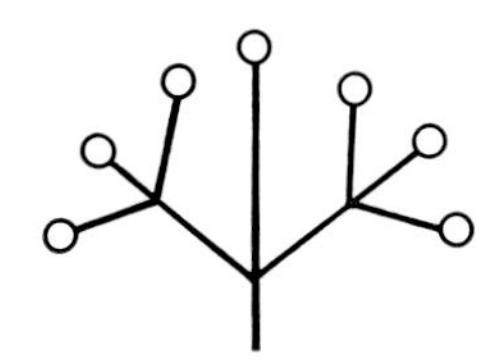

Dichlamydous: Said of a flower in which there are two whorls of perianth members.
Diclinous: Said of plants which possess two types of flower, one staminate, the other carpellate.
Dinkel: The bread or hexaploid wheats.
Dioecious: Possessing staminate and carpellate flowers but carried on separate plants.
Diploid: Twice the haploid chromosome number.
Disarticulation: The separation of individual parts or the whole spikelet of the grass plant.
Distichous: Having the leaves arranged in two diametrically opposed ranks.
Domatium: A specialized region on the lower surfaces of leaves, usually in the angles formed between the main and lateral veins, in which a characteristic microfauna or microflora can develop.
Double ridge: That stage in the development of the grass inflorescence when both the subtending bract and the inflorescence primordia have become visible.
Drupe: The fruit formed from a single carpel which consists of a fleshy mesocarp surrounded by a protective exocarp and having a hard stoney endocarp within which lies the seed. Sometimes more than one seed develops and each is surrounded by an endocarp. Aggregates of small drupes give a compound fruit, e.g. raspberry. In this case the fruit is called a **Drupel**.
Drupel: see **Drupe**.

Ecotype: A recognizably distinct form of a species found associated with a particular ecological situation. Free gene exchange can take place between ecotypes.
Einkorn: The small spelt or diploid wheats.
Emmer: The tetraploid wheats.
Endemic: Said of a species whose natural distribution is confined to a localized geographical region.
Endocarp: The innermost distinctive layer of the pericarp (fruit wall).
Endodermis: The innermost layer of cells of the cortex. In the root much modified by localized deposition of suberin and lignosuberin.
Endosperm: That tissue resulting from the development of the triple fusion nucleus after fertilization. May or may not persist until maturation of the seed is complete. A nutritive tissue for the developing embryo and/or young seedling.
Entomophily: The transferring of pollen from one flower to another by insects.
Epiblast: A flap-like mass of cells inserted on the embryonic axis of certain grasses opposite the point of attachment of the embryo to the scutellum.
Epicalyx: A second calyx close to the primary calyx of the flower and composed of sepal-like members (see Rosaceae).
Epigeal: Said of the germination of a seed when the cotyledons are brought above ground, i.e. the hypocotyl does elongate.
Epigynous: Said of a flower in which the sepals, petals and stamens are carried above the carpels; a flower in which the ovary is inferior.
Epiphyte: A plant which grows upon another plant, but not obtaining any nutrients from the association.
Erect: Upright.
Ergot: The sclerotium of the fungus *Claviceps purpurea* which replaces the grain in infected grass; the name also given to the disease. A hard horny dark-coloured object.
Essential oil: Volatile oils, usually fragrant, produced by plants in specialized glands or canals but sometimes not associated with any morphological or anatomical features. These are not triglycerides and are chemically terpenoids, extracted along with other compounds by steam distillation or hydrocarbon solvents. They are much used in flavourings and essences, hence the name.
Exocarp: The outer distinctive layer of the pericarp (fruit wall).
Exstipulate: Not having stipules.
Extrorse: Splitting along the outer surface in the anther to release pollen.
Extra-vaginal: Growth of an axillary bud through the surrounding leaf sheaths.
Eye: The colloquial name given to the bud on a tuber, particularly the potato tuber.
Eyebrow: The colloquial name given to the modified leaf subtending the eye (q.v.).
False septum (=**Replum**): An outgrowth from the placentae of the ovary of crucifers which partitions the ovary lumen into two.

Family: A distinct assemblage of genera with certain common features. A taxon of lower rank than an order.
Fasciation: Elaborate union of like members resulting in an abnormal appearance to the structures.
Fat: The ester formed between glycerol and the higher fatty acids; triglycerides. When the acids are highly unsaturated the fat is liquid at room temperature and is called an oil.
Fatuoid: A common oat with the features of *Avena fatua*. Thought to result from an aberrant chromosome situation.
Fermentation: The action of micro-organisms on organic substrates to produce carbon dioxide and low molecular weight alcohols, esters and/or acids.
Fibre: Any long thickened cell which possesses the capacity to withstand stress and strain. In the plant can act as mechanical support tissue and certain of these can be used in the production of thread and cloth. Other cells of similar morphology but not acting as support tissue are likewise exploited in the production of thread.
Filament: The stalk portion of the stamen.
Fixed oil: Triglycerides (q.v. **Fat**).
Flag leaf: The uppermost leaf on the stem of the grass plant.
Flavedo: The coloured portion of the pericarp of citrus fruits.
Fodder: The product derived from a forage crop.
Folioles: In a compound leaf those portions of the lamina directly attached to the petiole and which are not obviously leaflets.
Foliolules: Portion of lamina tissue attached to the stalks of leaflets of a compound leaf (see potato).
Forage: Any plant grown for its leaves and stems which are then consumed by domesticated animals.
Four-rowed: Said of barleys in which the spikelets are arranged in four discernible vertical rows.
Fruit: The structure derived from the carpels after the development of seeds.
Fuzz: The small brown contorted hairs on the seed coat of cotton.
Genome: The total chromosome complement of a gamete. Designated by capital letters to indicate the origin of the constituent sets.
Genotype: The genetic composition of an individual.
Genus: A taxon above the level of species and comprising a number of species with recognisable similarities.
Geotropism: The capacity for specific orientation within a gravitational field by selective growth within the responding organ. Negative geotropism—orientated so that growth is upwards (away from the Earth's centre); positive geotropism—orientated so that growth is downwards (towards Earth's centre).
Girdle: To remove a narrow strip of the outer layers of a woody stem such that it is encircled or almost so.
Glabrous: Smooth, without hairs.

Glume: A bracteate structure subtending the grass inflorescence. Sometimes used in other cases for a membranous bract.
Grain: The product of the grass flower which enters commerce.
Grain spirit: The distilled product of a yeast fermentation of ground ungerminated grain (cf. to malt spirit where the grain is allowed to germinate prior to yeast fermentation).
Gram stain: A staining procedure for differentiation of bacteria. It depends on the capacity of certain bacterial cell constituents to retain methyl violet after iodine treatment. To distinguish those which cannot from those which can safranin is often used as a counter stain. Gram +ve organism—stain blue; Gram −ve—stain red.
Grassland: A plant community in which the dominant species are grasses.
Green manure: Any plant grown for incorporation into soil to improve fertility; usually buried by ploughing.
Guard cell: The specialized cell of the epidermis which forms and controls the opening (pore) of the stoma.
Gynodioecious: Said of a species in which there are two classes of individuals; one with complete flowers, the other with only carpellate flowers.
Gynomonoecious: Said of a species in which the individuals possess both complete and carpellate flowers.
Gynophore: A distinct stalk on which the carpels are carried.
Haemagglutinin: Any substance which causes blood cells to agglutinate.
Halophyte: A plant which grows in soils with higher than average salt contents.
Haploid: The number of chromosomes possessed by the gametes, and half that of the somatic cells. The fundamental number of chromosomes observed in a polyploid series.
Hard wheat: Those bread wheats which can produce a flour that gives a tough dough for bread making; grown in regions where during grain filling the climate is dry.
Haustorium: A structure modified as an absorptive organ.
Heel end: The proximal end of a potato tuber—that end attached to the stolon.
Helophyte: An aquatic plant the leaves of which emerge above the water. At maturity there may not be free water and the rooting substrate may be mud.
Hermaphrodite: Possessing both male and female reproductive parts.
Hesperidium: The berry-like fruit of citrus trees.
Heterochlamydous: Said of a flower in which the outer protective parts are separated as distinct sepals and petals.
Heterosis: The greater more vigorous growth exhibited by hybrids when compared with their parents.
Hilum: (i) The central point around which starch grains develop. (ii) The scar remaining on the testa when the seed separates from its stalk (raphe).
Hinge cells: Longitudinal row of cells in the grass leaf which can change shape

as a result of turgor changes. Their movements cause the leaf blade to be expanded, or be rolled, or folded.

Husk: The fibrous covering of certain grains, e.g. oat: of little nutritional value and derived from the lemma and palea.

Hybrid vigour: see **Heterosis**.

Hydrophyte: A plant which grows either submerged or mainly in water.

Hypogeal: On germination where the cotyledons remain below ground. There is no extension of the hypocotyl.

Hypogynous: Having the sepals, petals and stamens inserted below the carpels. A flower with a superior ovary.

Imbricate: With overlapping margins.

Inbreeding: A breeding system where mating is between close relatives. In plants with bisexual flowers the maximum degree of inbreeding arises from self-fertilization. Long term inbreeding gives rise to genetically uniform progeny.

Induplicate: Folded or rolled inwards usually abruptly. Margins of adjacent members do not overlap. With the apex of the fold pointing downwards.

Inferior: The ovary in an epigynous flower (q.v.).

Inflorescence: The structure made by the aggregation of flowers on an axis. It may consist of single flower in which case the inflorescence is solitary.

Infructescence: The structure arising as a result of the production of fruits from the flowers of an inflorescence.

Intercalary meristem: A meristem (growth centre) intercalated between non-growing regions. (Contrasts with the normal meristems which are terminal.)

Intergenetic hybrid: A hybrid whose parents are individuals belonging to different genera.

Internode: That region of an axis lying between nodes.

Interspecific hybrid: A hybrid whose parents are individuals belonging to different species of the same genus.

Intravaginal: Said of an axillary shoot which grows within the leaf sheath of the subtending leaf.

Introrse: The splitting of an anther on its inner face to release pollen.

Isomerous: The floral whorls are each composed of equal numbers of members. Sometimes applied partially, e.g. stamens isomerous with petals meaning stamens equal in number to the petals.

Jorquette: A branch system in some tropical understorey trees where the lateral branches arise in clusters (false whorls) and are distinctly plagiotropic (see **Chupon**).

Keel: see **Carina**.

Kernel: The caryopsis of a grain which also includes the lemma and palea.

Kranz: Said of a plant with the following features: (a) a prominent bundle sheath; (b) the C_4 photosynthetic pathway; (c) generally a tropical or sub-tropical distribution; (d) a low CO_2 compensation point; (e) low discrimination against ^{13}C.

Laciniate: With the lamina deeply cut into segments.
Lamina: The terminal flat portion of a leaf.
Lanceolate: Having a lamina shaped like the head of a lance.
Latex: A milky liquid contained in canals within certain plants. On wounding the liquid exudes and may coagulate.
Leaf blade: see **Lamina**.
Leaf sheath: The basal region of some leaves developing to enclose younger leaves and stems and having a hollow cylindrical form.
Legume: A dry fruit, the product of one carpel and dehiscing along both the dorsal and ventral sutures; used to indicate that a plant belongs to the family Leguminosae.
Lemma: The lower of the bracteoles subtending the grass flower.
Leptomorph: A bamboo in which the rhizomes are thin and with noticeable internodes. Used in other species with similar rhizomatous systems.
Liana (**Lian**): A woody creeper of the tropical forest zones; often much contorted and with anomalous secondary thickening.
Ligule: Literally 'a tongue' and used to indicate tongue-like membranous structures found on leaves. The most characteristic is the one at the base of the adaxial surface of the grass leaf blade.
Lint: The long seed coat hairs of the cotton plant.
Loculus: The chamber of the ovary containing ovules; eventually a chamber in the fruit containing the seeds.
Lodging: The phenomenon seen in cereals where as a result of disease or adverse weather the culm is laid to the horizontal.
Lodicules: Small translucent–opaque structures in the grass flower, thought to be modified perianth members and possessed of the capacity to enlarge rapidly by an intake of water. They help to promote anthesis.
Lomentum: A linear fruit in which the individual seeds are separated by constrictions of the pericarp. At maturity the fruit breaks at the constrictions.
Maidens: A term used to describe vegetative reproduction units in their first year of growth.
Mericarp: The individual fruit of any dry compound fruit, but particularly that of a member of the Umbelliferae.
Mesocarp: The middle portion of the carpellary wall.
Mesocotyl: The middle portion of the embryonic axis lying between the hypocotyl and radicle, but used almost exclusively in the grass grain and none other. In this case it is defined as the part of the aerial embryonic axis attached and adjacent to the scutellum.
Micropyle: The small orifice where the integuments meet but do not fuse. This persists in the testa and can be seen in seeds.
Monadelphous: A set of stamens in a single bundle by connection of the filaments.
Monocarpic: Producing only one crop of fruits before dying.

Monochasium: A cymose inflorescence which develops a single system of branching, viz.

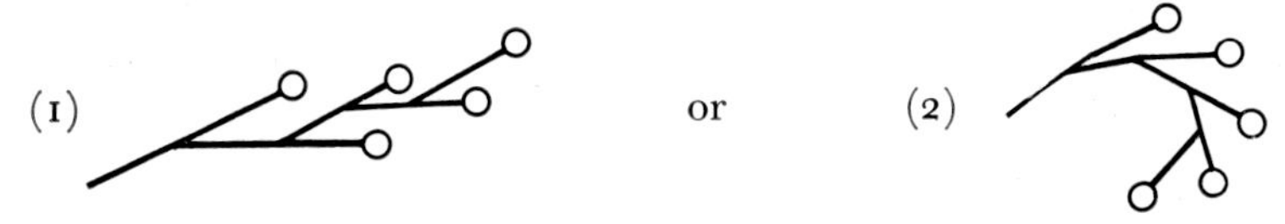

The first is the Cincinnus or scorpoid cyme, the second the bostryx or helicoid cyme.
Monochlamydous: Having a single protective layer, or perianth, to the flower.
Monoecious: Having staminate and carpellate flowers on the same plant.
Monogeneric: Said of any taxon higher than a genus which has but a single genus.
Monogerm: The condition in beet in which the infructescence has but a single fruit and hence a single seed.
Monophyletic: Having a single origin in geological time.
Monopodium: An axis which continues its growth by the activity of the terminal meristem.
Monospecific: Said of a genus which has one species.
Motor cell: see **Hinge Cell**, **Bulliform cell**.
Mucronate: Having the main vein projecting as an abrupt point.
Multifid: Divided at the tip into many parts.
Multipinnate: Divided pinnately into more than one set of pinnae (leaflets), so that there are at least primary and secondary systems of division.
Myrmecophily: The formal association between plants and a characteristic species of ant, e.g. in *Acacia*.
Nectary: A specialized structure secreting a sugary fluid (nectar) which may be in the flower (a floral nectary) or elsewhere (extrafloral).
Node: The point on an axis, frequently swollen, where the vascular system gives off traces to a leaf. A lateral bud is associated with the node in the axil formed between the leaf and the axis.
Nodule: Any round, or approximately so, protuberance.
Nut: A one-seed fruit in which the pericarp is woody or stoney (indehiscent).
Obcordate: Heart-shaped with the point of attachment at the pointed end.
Obdiplostemony: With two whorls of stamens the outer of which are opposite the petals and the inner opposite the sepals.
Obovate: An ovate leaf attached to the stalk by the narrower end.
Ontogeny: The course of development taken by an organism.
Open: With the petals arranged in a flower bud such that they do not touch and are separate.
Operculum: A specialized structure covering an opening through which seeds are dispersed.
Order: A taxon comprising a group of related families.

Orthotropous: Said of an ovule which is upright with the micropyle directly opposite the point of attachment of the raphe to the placenta.
Out-breeding: The condition in sexual reproduction of the parents not being genetically related.
Pachymorphic: Said of a rhizome system composed of axes with short thick internodes (see **Bamboos**).
Pale: A general term for the two bracteoles subtending the grass flower.
Palea: The innermost of the two bracteoles subtending the grass flower.
Palmate: Having the veins of a leaf radiating from a common origin where the stalk meets the blade. Applied also to compound leaves and to the form of lobing in a simple leaf where these follow the pattern of palmate veins.
Palmatifid: Having a palmate leaf blade divided deeply between the veins.
Panicle: The name given to a grass head in which the inflorescences (spikelets) are carried on stalks. The arrangement is various and may be open in which case the stalks from the main stem (rachis) are long and few and may branch, or contracted where the stalks are short and many.
Panmixis: With the opportunity of free recombination of characters within a population (cf. **Apomixis**).
Papilionaceous: Like a butterfly.
Parthenocarpy: The production of fruits in the absence of fertilization.
Passage cell: A specialized cell of the endodermis without secondary wall production lying opposite the protoxylem.
Pedicel: The diminutive form of peduncle; a small peduncle.
Peduncle: The flower stalk.
Peepers: Small suckers of banana just erupting through the soil surface and without expanded leaves.
Pellucid: Translucent, watery.
Pentamerous: Having the flower parts in fives or multiples of five.
Pentaploid: Having five times the basic number of chromosomes.
Pepo: The fleshy berry-like fruit produced by members of the Cucurbitaceae.
Perennial: A plant which persists for more than two years.
Pericarp: The fruit covering derived from the ovary wall.
Perigynous: Having the sepals, petals and stamens inserted on the torus on a plane mid-way on the ovary. With neither an inferior nor superior ovary.
Perimedulla: That part of the medulla (pith) on the periphery and adjacent to the vascular tissues.
Petal: The organ of the distinctive second layer of the perianth of a flower which is heterochlamydous and least like a leaf. Usually coloured and delicate.
Petiole: The stalk of a leaf.
Petiolule: A second-order petiole hence the stalk of a leaflet.
Phellem: Cork.
Phelloderm: Layers of parenchymatous cells produced internally by the cork cambium.
Phellogen: A layer of meristematic cells on the periphery of an axis which

produces mainly the cells of the bark or other protective layer—the cork cambium.

Phenotype: The eventual expression of the genotype of an individual interacting with the environment.

Photomorphogenetic: Said of a growth process the development of which is controlled by light.

Photoperiod: A characteristic daylength which occasions specific photomorphogenetic (q.v.) events in the life of a plant.

Phototropic: A directional movement towards (+ve) or away from (−ve) light brought about by unilateral growth.

Piliferous layer: The outer root-hair bearing layer of cells in a root.

Pinna: A small segment of a divided lamina.

Pinnate: Having a leaf with a main vein from which at intervals along it, side veins arise. A leaf blade divided between such veins.

Pinnatifid: A pinnately veined leaf with the lamina deeply cut between the lateral veins.

Placenta: The ovule-bearing tissue of the carpellary wall.

Plagiotropic: Having a growth habit which is not vertical. More or less horizontal.

Ploidy: The level of multiplication of the basic chromosome number in a polyploid organism.

Plumule: That part of the embryonic axis which eventually gives rise to the stem and transition zones of the main axis.

Pole bean: Any bean crop which exhibits monopodial growth and which for maximum yield, has to be grown in a trained fashion.

Polyarch: Having a large number of individual primary stele initials. Literally should refer to the procambial strand initials.

Polycarpic: With the capacity to flower and fruit an indefinite number of times.

Polyembryony: The state where a seed has more than one embryo.

Polyphyletic: Said of a species (or other taxon) which is considered to have arisen more than once in geological time.

Pome: The fleshy 'fruit' of apples pears and quinces. Since the ripe structure contains material derived from the axis it is botanically speaking more complex than a true fruit.

Primordium: The first recognizable, but undifferentiated, stage in the development of an organ.

Procumbent: Lying on the ground but not rooting at the nodes.

Propagule: That part of a plant employed in its propagation (by vegetative means). Literally, a seed can also be classed as a propagule.

Prophyll: The oldest leaf of an axillary bud, often modified. In grasses the prophyll is tubular with a closed tip and is incapable, in the light, of much extension.

Prostrate: Lying flat on the ground.

Protandry: In a flower having the stamens ripe and shedding pollen before the stigmas are receptive.
Protogyny: In a flower having the stigmas receptive before the stamens are ripe and shedding pollen.
Pseudospike: The head of ryegrass in which opposite sessile spikelets are arranged radially and not tangentially as in a true spike.
Pseudostem: The upright structure formed by a large number of successive sheathing leaf bases, e.g. banana.
Pubescent: Possessing fine downy hairs.
Pulse: An edible seed of any leguminous plant.
Pulvinus: A cushion-like swelling at the base of a leaf. Leaf attitude can be altered by changes in turgor within pulvinar tissue.
Race: A group of individuals of a species probably of common genetic origin and recognisably distinct from other such groups. Could be equated with variety.
Raceme: An indeterminate inflorescence in which the youngest flowers are at the apex.
Rachilla: A branch from a rachis (literally a small rachis).
Rachis: That part of the grass stem (culm) which carried inflorescences.
Radial: Said of anything which has an obvious direction from a centre towards the periphery of an imaginary circle. Lying parallel or with the long axis in the same direction as a radius of such a circle.
Radicle: That part of the embryonic axis which gives rise to the main root system of a plant.
Ratoon: Sprouts or shoots arising from the basal or subterranean part of a plant after it has been cropped.
Receptacle: That terminal part of an axis which carries the flower parts, or in the case of a composite head the florets.
Reduplicate: Folded in such a way that the apex of the fold is pointing upwards.
Replum: False septum.
Rhizome: A subterranean stem producing adventitious roots at the node and here and there aerial shoots. May be monopodial, in which case the terminal bud is always below ground, or sympodial in which case the terminal bud becomes aerial and growth of the rhizome system is continued by a lateral bud.
Rootstock: In a graft system that part possessing the roots and basal portion of the stem.
Rose end: The apical region of a potato tuber.
Rosette: A plant form in which internodes are very short and the leaves are arranged close together as a circular cluster.
Ruminate: Much folded and irregular in appearance, as though chewed.
Runner: A slender prostrate stem with long internodes and rooting at the nodes so establishing new plants from the axillary buds once the internode region has rotted.

Saponins: Plant constituents, usually steroidal glycosides, with soap-like properties inasmuch as their solution foams vigorously on shaking; can act like detergents.
Scale leaf: A structure with the origin of a leaf but not possessing the characteristic form of that organ; usually small, perhaps green but not regularly so, and often horny.
Scion: That part of a graft system which forms the aerial portion of the plant. Attached to the rootstock (q.v.) by a graft union.
Sclerophyllous: Possessing much woody tissue, either as woody fibres or other cells with lignified walls.
Scutellum: The shield-shaped mass to which the embryo is attached in the grass caryopsis.
Secund: Being arranged on one side. Used especially with regard to paniculate heads.
Seed: The structure derived from the ovule after fertilization of the egg. A reproductive structure usually resistant to adverse conditions.
Self-fertile: Able to form zygotes by union of gametes produced by the same plant.
Self-incompatibility: Inability to produce zygotes by union of gametes produced by same plant, hence on self-pollination seeds are not formed.
Self-pollination: Transfer of pollen from the stamens to the stigma of the same flower or to another flower of the same individual.
Seminal root: A root arising from a primordium recognizable on the embryonic axis.
Sepal: The organ of the outer of the two protective layers in a heterochalamydous flower. That structure of the flower most nearly like a leaf.
Serrate: Toothed like a saw.
Sessile: Attached directly to the axis, there not being evidence of a stalk.
Sett: (i) A sucker shoot used to vegetatively reproduce a plant. (ii) A portion of a rhizome containing a bud and which can be used in the same way as (i).
Silage: The product obtained by a controlled anaerobic fermentation of green moist plant parts. Used as an animal foodstuff.
Silica cell: Specialized cells of plants in which silica is deposited as a quasi-crystalline mass.
Silicula: A cruciferous fruit as broad as it is long.
Siliqua: A cruciferous fruit much longer than broad and usually with a beak.
Silks: The colloquial term used to describe the mass of long styles found on an ear of maize.
Single ridge: That stage in the transition from the vegetative to reproductive apex of the grass where the bract primordia are just recognisable.
Six-rowed: Said of barley heads where the lateral spikelets do not overlap and so six distinct longitudinal rows of grains are seen.
Slip: A lateral shoot removed to be used in vegetatively propagating a plant.
Soft fruit: A commercial term to indicate the fruits of strawberry, raspberry,

gooseberry, blackcurrant, and generally the berries obtained from small shrubs and perennial herbs.

Soft wheat: Any wheat which gives a flour more suited to the production of biscuits and pasta, but not suited for baking bread (cf. hard wheat).

Spadix: A racemose inflorescence with a fleshy or thickened axis sometimes with flowers embedded within it.

Spathe: Large, often showy, bracts enclosing a spadix (q.v.).

Spathulate: Shaped like a spoon.

Species: A taxon the definition of which causes some controversy—fundamentally, a group of individuals potentially able to exchange genes freely.

Spelt: The character seen in grass spikes where the rachis breaks along its length to separate individual spikelets

Speltoid: Said of certain aberrant individuals of bread wheat which exhibit *inter alia* the spelt character. Thought to be due to an abnormal cytological situation.

Spiculate: Covered with small spines.

Spike: A racemose inflorescence in which the flowers are sessile or nearly so.

Spike-like panicle: A grass head which is fundamentally a panicle but because of contraction of the rachillas it appears like a spike.

Spikelet: The true inflorescence of the grass containing 1–15 flowers subtended by two bracts (=glumes).

Spur bearer: Said of any of the fruit trees which carry fruits on spur shoots.

Spur shoot: Small dwarf lateral shoots produced by many trees; nearly always associated with the production of flowers.

Stamen: One of the essential parts of the flower situated distal to the innermost part of the perianth. It produces pollen and consists of a stalk or filament and a head or anther.

Staminate: Said of a flower in which the only essential parts are stamens.

Staminode: A sterile structure replacing a stamen; of various forms.

Standard: The median posterior petal of the leguminous flower.

Steckling: A beet plant which overwinters in the reproductive state with long stick-like stems.

Stele: Those tissues of the axis involved in the long distance transport of materials, and their adjacent parenchymatous cells.

Stipules: A pair of leafy structures often found at the base of a leaf where it joins the stem; may be reduced to hairs, and can be much modified.

Stolon: A horizontal stem produced at ground level and growing along the ground rooting at the nodes.

Stone fruit: Those large drupes of the Rosaceous trees grown as food crops, e.g. cherry, plum, peach.

Strain: A commercial term used to denote a recognisable phenotype and hence possibly genotype. Equivalent to race (q.v.).

Strand: That part of the land mass which borders the shore.

Strangles: A condition of some root crops, especially beet, where the peripheral

tissues of the young seedling are desiccated so giving a pinched appearance to the hypocotyl and lower part of the stem.
Striae: Parallel rows of lines, or grooves.
Strut root: Adventitious roots produced at the basal nodes of many large grasses. They grow at an angle, like a guy rope, to secure the stem by becoming anchored in the soil.
Stylopodium: The glandular disc surmounting the receptacle in Umbelliferous plants and on which the styles are carried.
Sub-species: A taxon of lower status than species but ranking above variety.
Subsidiary cell: Those epidermal cells adjacent to the guard cell and involved in stomatal movement just as much as those cells which visibly change shape.
Sucker: A quick-growing shoot, usually sappy, arising near the base of the stem or on the subterranean parts close to the stem.
Superior: Said of a flower in which the gynoecium is placed above all other parts.
Suture: The line where two parts meet.
Sward: The closed plant community formed by grasses and usually maintained by grazing or cutting.
Sword leaf: Any upright unopened leaf that is shaped like a sword (see Palms).
Sword shoot: Any upright large unopened bud shaped like a sword (see Banana).
Symbiosis: The association between two organisms for their mutual benefit.
Sympodium: A growth pattern of an axis system in which continued growth is maintained by the successive activites of lower order branches. Any one branch grows for a limited time then a bud of it takes over the growth of the plant. After a period this pattern is repeated.
Syncarp: An aggregate fruit formed by the coalescence of adjacent fruits to give a large coherent mass. The axis carrying the fruits may become involved e.g. pineapple.
Syncarpous: The condition in a flower where the individual carpels fuse to give a single ovary.
Syngenesious: Having stamens in bunches as a result of fusion of anthers.
Tapping: The obtaining of any liquid exudate of plants by slashing, cutting or boring a woody stem, or other robust organ.
Tassel: The name given to the large open terminal panicles of certain large panicoid grasses, particularly maize.
Taxon: Any recognizably distinct grouping which can be used to indicate the systematic and taxonomic relationships of organisms.
Tendrils: Organs adapted for attaching to substrates by means of coiling. May be derived from many structures, e.g. stipules, leaflets, branches, etc.
Tepal: a non-committal term applied to the members of the perianth not indicating whether they are like sepals or petals.
Teratological: Said of any organ exhibiting gross abnormality or monstrosity.

Ternate: A leaf blade in three parts.
Testa: The covering of a seed derived from the integuments of the ovule.
Tetradynamous: Having four long and two short stamens.
Tetramerous: Having the flower parts in fours or multiples of four.
Tetraploid: With the cells having four times the basic number of chromosomes.
Theca: The sac containing the pollen (generally any sac-like structure containing spores, etc.).
Tillers: The axillary branches of a grass.
Tip bearer: Those top and stone fruits in which the branches terminate in regions with the characters of a dwarf shoot, hence flowering and fruiting at the ends of such branches.
Top fruit: A commercial term for apples and for pears.
Torus: That part of the axis on which arise the flower primordia—the receptacle.
Tribe: A taxon consisting of a number of related genera.
Trichome: A hair developing from the epidermis.
Trifoliate: Having three leaflets.
Triploid: Having three times the basic number of chromosomes.
Truss: The term used by commercial growers for any infructescence carrying a large number of discrete fruits, e.g. tomato.
Two-rowed: Said of a barley in which only the median spikelets produce grains so that on the mature head there are only two longitudinal rows of grain.
Umbel: A racemose inflorescence in which all the peduncles arise at the same point on the axis and as a result of being of various lengths the individual flowers are brought to lie in a single plane.
Umbellule: A small umbel carried on a stalk which is itself part of an umbellate system; a secondary umbel.
Unifoliate: Having a simple single leaf blade.
Univalent: A chromosome left unpaired at meiosis.
Urnicate: Shaped like an urn.
Valvate: Having the petals folded in the bud such that the margins meet but do not overlap.
Variety: (i) A taxon below the level of sub-species. (ii) A term used by seedsmen to indicate a distinct strain.
Vermifuge: A substance which will expel parasitic worms from animals.
Vernalization: The exposing of plants to low temperature to bring about changes that will permit flowering.
Vexillum: see **Standard**.
Vittae: A canal into which is secreted oil, found in the Umbelliferae and giving their fruit a striped appearance.
Ware: The category of produce acceptable for marketing for human consumption.
Water table: The depth below which the ground is saturated with water.
Whorl: Having the parts arranged as in a circle.

Wings: see **Alae**.
Xeromorphic: Said of morphological characters thought to confer upon a plant the ability to withstand shortage of water.
Xerophyte: A plant able to grow under conditions where available water is limited.
Zero grazing: The agricultural practice of keeping animals indoors and carrying to them green plant parts which, under open or managed grazing, the animals would eat directly from the growing plant.
Zygomorphic: Said of a flower in which there is only one plane, usually median, of symmetry (=irregular).

Species List

Abelmoschus esculentus (L.) Moench.
Abrus precatorius L.
Aegilops comosa Sibth. et Sm.
  speltoides Tausch
  squarrosa L.
Aethusa cynapium L.
Agrostis canina L. subsp. *canina*
  canina L. subsp. *montana* Hartm.
  gigantea Roth.
  stolonifera L.
  tenuis Sibth.
Aleurites fordii Hemsl.
  montana (Lour.) Wils.
Allium ampeloprasum L.
  cepa L.
  chinense Don.
  fistulosum L.
  sativum L.
  schoenoprasum L.
Alopecurus pratensis L.
Ammophila arenaria (L.) Link
Ananas comosus (L.) Merr.
Anthyllis vulneraria L.
Apium graveolens L.
Arachis hypogea L.
Armoracia lapathifolia Gilib.
Arundinaria tecta (Walt.) Muhl.
Atriplex patula L.
Avena brevis Roth.
  fatua L.

Avena—contd.
  nuda L.
  sativa L.
  sterilis L.
  strigosa Schreb.
Axonopus compressus (Swartz) Beauv.

Bambusa vulgaris Schrad. ex Wendland
Beat vulgaris L. ssp. *vulgaris*
  ssp. *maritima* (L.) Thell.
Brachiaria brizantha (Hochst.) Stapf.
  decumbens Stapf.
  mutica (Forsk.) Stapf.
Brassica campestris L.
  carinata A. Braun
  chinensis L.
  juncea (L.) Czern. et Coss.
  napobrassica (L.) Mill.
  napus L.
  nigra Koch
  oleracea L.
  pekinensis Rupr.
Bromus inermis Leyss
  unioloides HBK

Cajanus cajan (L.) Millsp.
Camellia sinensis (L.) O. Kuntze

Canavalia ensiformis (L.) D.C.
gladiata (Jacq.) D.C.
Capsicum annuum L.
frutescens L.
Cercis siliquastrum L.
Chenopodium album L.
ambrosoides L.
quinoa Willh.
Cicer arietinum L.
Cicuta virosa L.
Citrullus lanatus (Thunb.) Mansf.
Citrus aurantifolia (Christm.) Swing.
aurantium L.
grandis (L.) Osbeck
limon (L.) Burm. f.
medica L.
paradisi Macf.
reticulata Blanco
sinensis (L.) Osbeck
Claviceps purpurea (Fr.) Tul.
Cocos nucifera L.
Coffea arabica L.
canephora Pume ex Frochner
Coix lachryma-jobi L.
Conium maculatum L.
Corylus avellana L.
Croton tiglium L.
Cucumis melo L.
sativus L.
Cucurbita maxima Duch. ex Lam.
mixta Pang.
moschata (Duch. ex Lam.) Duch. ex Poir.
pepo L.
Cynosurus cristatus L.

Dactylis glomerata L.
Daucus carota L.
Dendrocalamus giganteus Munro
Desmodium gyrans D.C.
Digitaria decumbens Stent.
Dolichos uniflorus Lam.

Echinochloa frumentacea (Roxb.) Link.
Elaeis guineensis Jacq.
Elymus arenarius L.
Euchlaena mexicana Schrad.

Faba vulgaris Moench.
Festuca arundinacea Schrad.
ovina L.
pratensis Huds.
rubra L. subsp. *rubra*
rubra L. subsp. *commutata* Gaud.
Fragaria ananassa Duchesne
chiloensis Duchesne
vesca L.
virginiana Duchesne

Glycine max (L.) Merr.
ussuriensis Regel. et Maack
Gossypium arboreum L.
barbadense L.
herbaceum L.
hirsutum L.
tomentosum Nutt. ex Seem.

Hevea brasilensis (Willd. ex Adr. de Juss.) Muell-Ang.
Hibiscus cannabinus L.
esculentus L.
Hordeum distichon L.
agriocrithum Aberg.
spontaneum C. Koch
vulgare L. emend.
Hyparrhenia rufa (Nees.) Stapf.

Ischaemum indicum (Houtt.) Merr.

Lablab niger Medik.
Lens esculenta Moench.

Lolium multiflorum Lam.
perenne L.
rigidum
temulentum L.
Lotus corniculatus L.
pedunculatus Cov.
Lycopersicum esculentum Mill.
pimpinellifolium (Jusl.) Mill.

Malus pumilis Mill.
sylvestris Mill.
Manihot esculenta Crantz.
Medicago lupulina L.
sativa L.
Melilotus alba Desr.
Melinis minutiflora Beauv.
Mimosa pudica L.
Musa acuminata Colla
balbisiana Colla
textilis Nee.

Nicotiana rustica L.
tabacum L.

Oenanthe crocata L.
Onobrychis sativa Lam.
Ornithopus sativus Brot.
Oryza glaberrima Steud.
perennis Moench emend Sampath
sativa L.

Pachyrhizus erosus (L.) Urban
tuberosus (Lam.) Spreng.
Panicum maximum Jacq.
miliaceum L.
Paspalum conjugatum Berg.
dilatatum Poir.
scrobiculatum L.
Pennisetum clandestinum Hochst ex Chior.
purpureum Schum.
typhoides (Burm. f.) Stapf et Hubbard
Peucedanum sativum Benth.
Phalaris arundinacea L.
canariensis L.
tuberosum L.
Phaseolus aconitifolius Jacq.
angularis (Willd.) Wight
aureus Roxb.
coccineus L.
lunatus L.
mungo L.
vulgaris L.
Phleum nodosum L.
pratense L.
Phoenix dactylifera L.
Phytophthora parasitica Dastur.
Piper nigrum L.
Pisum arvense L.
sativum L.
Poa compressa L.
pratensis L.
trivialis L.
Poncirus trifoliata L.
Prunus armeniaca L.
cerasus L.
communis Batsch.
domestica L.
persica (L.) Batsch.
Psophocarpus tetragonobolus (L.) D.C.
Pyrus communis L.
malus L.

Raphanus sativus L.
Ricinus communis L.
Rhizobium japonicum (Kirch.) Buch.
leguminosarum Frank emend. Baldwin et Fred
lupini (Schr.) Ecktr. *et al.*
meliloti Dang.

Rhizobium—contd.
phaseoli Dang.
trifolii Dang.
Rubia tinctorum L.
Rubus chaemaemorus L.
fruticosus L.
idaeus L.
saxatilis L.
ursinus Charn. et Schlecht.

Saccharum barberi Jesweit
officinarum L.
sinense Roxb.
spontaneum L.
Secale anatolica Boiss.
cereale L.
montanum Guss.
segetale Zuk.
Setaria italica (L.) Beauv.
sphacelata (Schum.) Stapf et Hubbard
Solanum melongena L.
nigrum L.
tuberosum L. ssp. *tuberosum*
ssp. *andigena*
Sorghum bicolor L.
halepense (L.) Pers.
Spinacia oleracea L.
Stizolobium deeringianum Bort.
Hasojoo Piper et Tracey
Stylosanthes guyanensis (Aubl.) Swartz
humilis H.B.K.

Themeda triandra Forsk.
Theobroma cacao L.
Trifolium alexandrinum L.
dubium Sibth.
hybridum L.
Trifolium—contd.
incarnatum L.
pratense L.
subterraneum L.
repens L.
Tripsacum dactyloides L.
Triticum aestivum L.
boeticum Boiss.
carthlicum Nevski
compactum Host.
dicoccoides Koern.
dicoccum Schubb.
durum Desf.
macha Dek. et Men.
monococcum L.
polonicum L.
spelta L.
sphaerococcum Perc.
timopheevi Zukov.
turanicum Jakubz.
turgidum L.
vavilovi Tuman.

Urophylyctis alfalfae (Lagerh.) Magnus.

Vetiveria zizanioides (L.) Nene
Vicia faba L.
sativa L.
Vigna sesquipedalis (L.) Fruw.
sinensis (L.) Savi ex Hassk.
unguiculata (L.) Walp.
Voandzeia subterranea (L.) Thouars

Zea mays L.
mexicana Reeves et Mangelsd.

Index

D

E

F

U

V

W

X

Y

Z